The Black Technical Object: On Machine Learning and the Aspiration of Black Being

ON THE ANTIPOLITICAL

The Black Technical Object

On Machine Learning and the Aspiration of Black Being

Ramon Amaro

Sternberg Press

CONTENTS

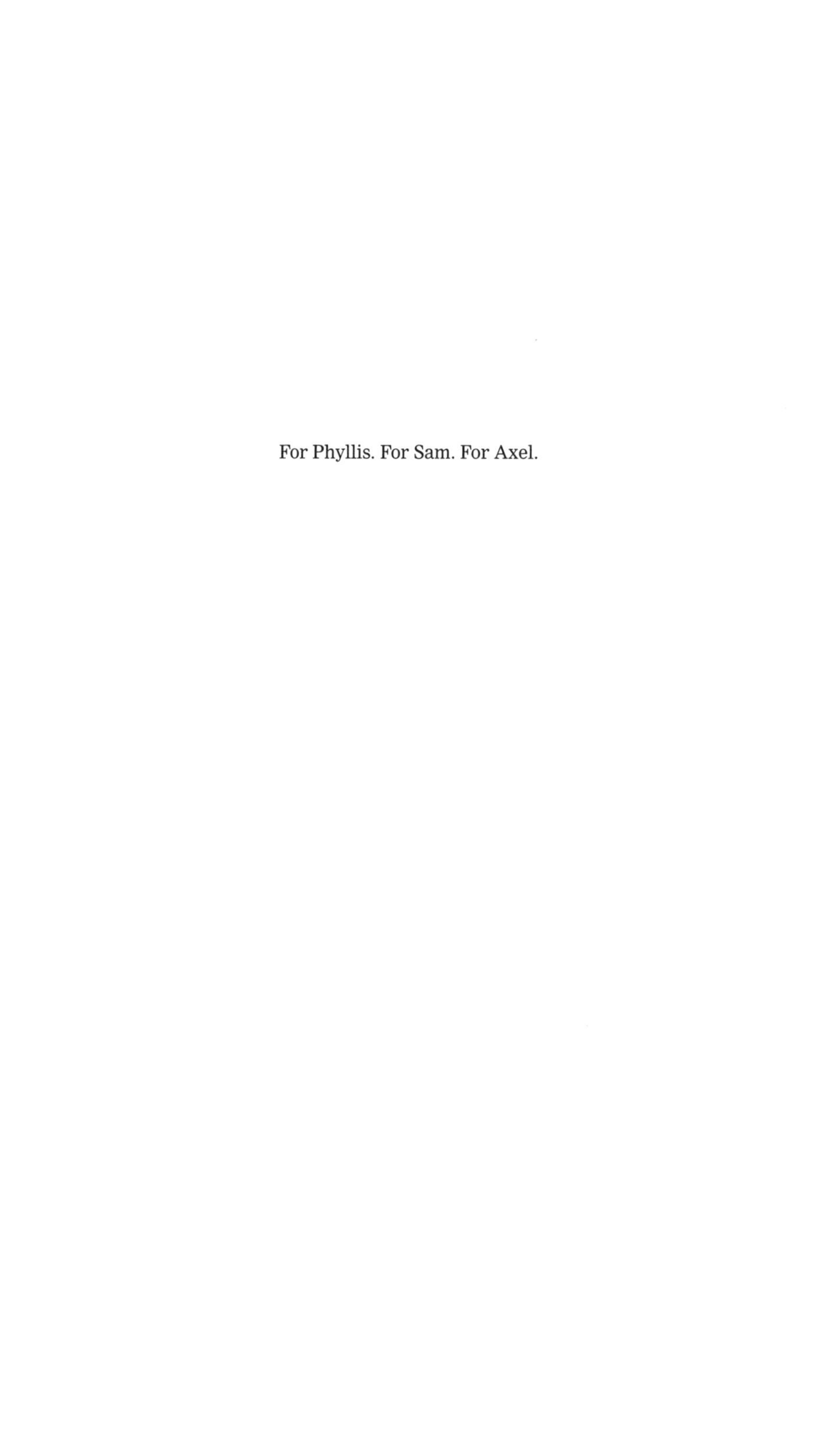

For Phyllis. For Sam. For Axel.

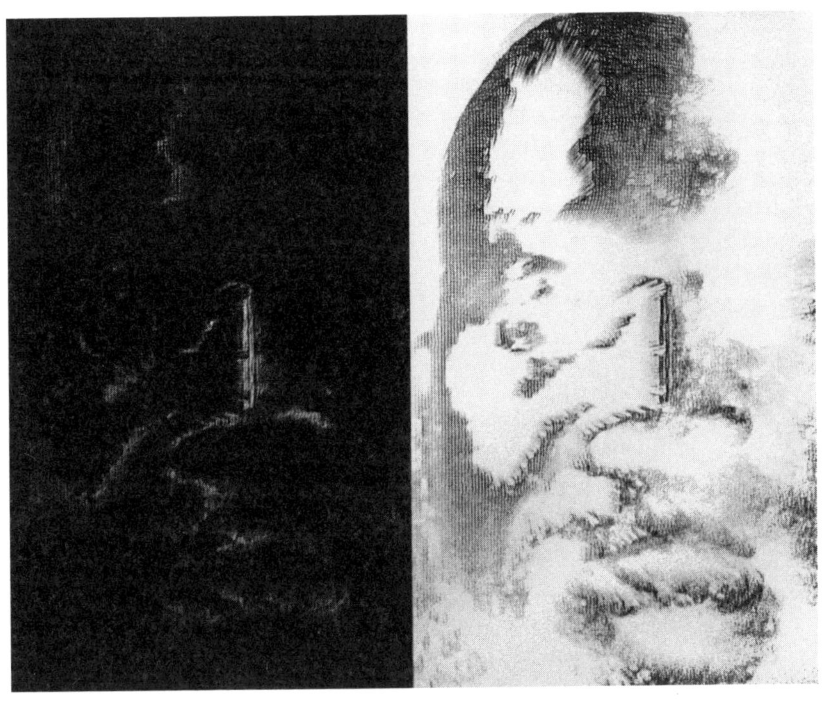

Ryo Ikeshiro
Ethnic Diversity in Sites of Cultural Activity, 2014–ongoing
Image from performance at Chisenhale Art Place,
London, 2015

INTRODUCTION
Black Psychic and Machinic Alienation

This book is informed by the incompatible relation between machine learning, data, and race. As an exploratory work, it is motivated by the desire to raise awareness of machine learning as that which is organized around the presupposition of racial categorization. It responds to an inadequate understanding of machine learning and Black existence. From this inadequacy, we have, in the first instance, become aware of the veritable attitude toward machine learning and Black existence as being in and of themselves. A gap then manifests in the tendency to establish an oppositional ground between Black existence and machines. These attitudes presume a determinacy, oversimplifying notions of what it means to be human and what it means to be machine. Thus far for the human, the function of this ensemble is to alert one to the recurrent cycles of racialized information.

The evolution of machine learning is in fact conditioned by a set of atmospheric conditions that have attempted to organize both humans and technical objects around assumptions of race. Machine learning provides, in a sense, the assumptions on which rest the epistemic claims of racial objectivity; however, its technology is preconditioned by a process of human relations that has already conceived of Black beings as objects among other objects, following Frantz Fanon. If this is the case, the fact of Blackness as an object of being enters into relation with the technical object as always already externally fragmented and thereby readied for algorithmic reassembly through the white imaginary.

This epistemic process has aroused a sense of danger evident in the harboring of human alienation and the self-negation of Black existence caused by the machine. For the algorithmic, however, this entanglement is conceived of as the willful perversion of a recurrent human reality. While these states of being constitute an offensive system of techno-human inequity, they furthermore presume an exclusivity over the value of life in either defense of the technical or the preservation of humankind. The relation ignores the fact of a human reality fully incorporated into technical being, which must be realized through a reciprocal value system with a common ground of existence. Here, Black existence, as milieu and as the basis for affirmative livability, arrives as a continuation of the historical, and as the rupture between present duress and future aspiration. The algorithmic, on the other hand, is granted the true and full nature of its own existence, set free from its devaluation as mediator between user/engineer or consumer/capital and its general evolution as the interpreter for a succession of negating human values.

This becoming techno-human, becoming of the Black technical object, is envisioned at the level of the algorithm, as a mode of access to the historicity of machine learning's becoming thus far. The mode of access to machine learning is both major and minor. It is minor in the sense that the technology is conditioned by a knowledge of its tools and instruments: its mathematical undercarriages, its technical capabilities, the devices that house them, and their actionable objectives—all which have turned humans into the bearers of primitive instruments according to a distinctive set of actions and behaviors. A major mode of access is presupposed in our becoming aware of the ways in which machine learning functions, at the level of the algorithm, to gain from its evolutionary relationship with other diverse modes of technical knowledge (statistics, computer science, pattern recognition, artificial intelligence) and the various value judgments made by those regarding their power. However, the notion of Black existence materializes in the system on an equal basis, whereby the minor mode of access to the appropriate conditions of Blackness is centered on an awareness of the conditions that underline our presumed capabilities, devised both internally and externally according to an expected set of actions and behaviors regarded by those in power as a determinant set of values. But what if, under the irreversible circumstances of the white protean imaginary, the Black object was to build a new relation with the machinic? What if this Black technical object was to interact with the logics of machine learning beyond the desire for recognition and reinforcement of its existing rudimentary operations? What if we, the Black technical object, were to travel through the algorithmic as that which enacts its own form of reason, to arrive at a self-actualization—or what Fanon calls an "effective disalienation"?[1] What if, as Stefano Harney and Fred Moten argue, the Black technical object were to take a right of refusal to racial perception, and aspire to be that which is out of reach of the negating factors of race?

This book intends to provoke an answer to these questions by providing both minor and major access to techno-human reality. This provocation is motivated by a desire to raise awareness of the mode of existence of machine learning and the mode of existence of Black being in order to arrive at an alternative ground that enables the full potential of Blackness and the technical object. It seeks to become aware of the Black technical object's becoming. This book is therefore as speculative as it is instructional, in that it is grounded in the reality of data, machine learning technology, and conditions that inform the negating perception of Black being. As such, it is a circular narration of these seemingly disjointed lines of inquiry in an attempt to establish a prognosis of these realities. Accordingly, it seeks to establish a foundation of these lines of thought that must be readily understood if we are to achieve its aims. By situating this problematic within the metaphysics of Black being, in conversation with machine learning's technical and theoretical discourses, this work is a point of departure from which we can consider the future of machine learning and the future of racial beings by examining the vast and historical conditions that inform our present understanding of race and technology. In my attempt to see what machine learning is—both functionally and theoretically—I am also seeking to understand what technology might allow for in the reconfiguration of Black being. This inquiry into machine learning and the metaphysics of Black being "let me do both," as Octavia Butler says of science fiction. It lets me "look into science and stick my nose in everywhere."[2]

1 Frantz Fanon, *Black Skin, White Masks*, trans. Charles Lam Markmann (London: Pluto Press, 2008), 4.

2 John C. Snider, "Interview: Octavia E. Butler," in *Conversations with Octavia Butler*, ed. Conseula Francis (Jackson: University Press of Mississippi, 2010), 217.

We Believe in Data

We are a social body that feeds on a junk pile of fragmented experiences. We are brought into visibility by this insatiable need to for the algorithmic. We rely on this empty vessel of culture to satisfy our overindulgence in a world that we made in our own image. This image not only illuminates a dissatisfaction with the violences the West has projected onto the world, but is also a working model of the epistemic deficiencies of logical positivism. Here quality of life is characterized as livable, or worthy to be lived, only inasmuch as the basis of life can be characterized by the assertion of administrative code and belief. At the same time, a supernatural belief of this vague idea of data—in which the West has placed its confidence—loses its efficacy when faced with the very real notion of a life outside of its own obsession with social administration.

We, particularly in the West, would like nothing more than to release this neurosis by filling our emptiness with aspirations of continued global ascendance, built on the systemic fragmentation of difference. That is to say, the recursive features of our Western imaginary are reliant on the reinstatement of logical "truths" made actionable by systems of categorization. This act of self-validation is activated by the substance of race, which is a particular kind of social constitution that restores faith in providence while displacing judgment toward the racialized Other. The adoption of this imaginative creation is executed by the management of knowledge-based procedures, exercised in the everyday workings of scientific principles. Under the organization of these principles, technology stands in as a beacon of progress where dominance itself is metaphor for a devout commitment to the collection of facts about circumstances that exist or events that have already occurred. But we already know that a concept whose truth can be proved is no more than the regulation of the experiential and, therefore, the aspirational characteristics of life.

It is no surprise that machine learning has become the vessel through which discriminate human relationships are understood. Machine learning is a reawakening of Western attempts at mechanizing culture for its own aims. This is achieved through the conversion of contingent social compositions into judicial principles of assertion. That is, the declaration of authority is self-validating in its systemic appropriation of life by means of taxonomic division, particularly in relation to the categories of race. According to Paul Gilroy, racial taxonomies work to reinforce racial fragmentation. Although he argues that the flattening of cultural difference by means of classification is far less valuable to commercial markets than racial alterity, he is concerned with the "raw untrained perception dwelling in [a] body" that has conceded to the possibility of race.[3] As he remarks, "The human sensorium has had to be educated to the appreciation of racial differences. When it comes to the visualization of discrete racial groups, a great deal of fine-tuning has been required."[4] Gilroy's attention to the sensorium brings our conscious awareness of racial taxonomy into question as a problem of technological optimism. "The call of racial being," he argues, "has been weakened by [...] the idea that the body is nothing more than an incidental moment in the transmission of code and information, by its openness to the new imaging technologies, and by the loss of mortality as a horizon against which life is to be lived."[5] Although it is unclear whether what is considered strength in this case is in fact a reciprocal causality between one's obedience to the universal imaginary and the true meaning of life, there is balance to be found in building an awareness of structures that exist as grounds for personal fulfillment. With the horizon in tow, the notion of information allows technology to be interpreted in various ways, and by means of a linear succession of elemental events that are assigned value according to the laws of life's conservation. Nonetheless, the notion of death as the tool bearer for a life well lived is a disingenuous nostalgia masked as a basic doctrine of signification.

---

3 Paul Gilroy, *Against Race: Imagining Political Culture Beyond the Color Line* (Cambridge, MA: Belknap Press, 2000), 52.

4 Gilroy, 42.

5 Gilroy, 36

The hypothesis consists of the employment of a doctrine of life that is assumed lost in its relation to the world.

The pessimistic and dramatic usurping of technological being—as if there is a neutrality and subsequent power between the artificial and the sacredness of human life—does not deny the existence of an elaborate shift in aesthetic activity. The imaging and imagining of race in machine learning algorithms make apparent how an overdependence on technological solution produces an anxiety resulting from the frustration of its progress. After all, it might indeed take a prophet to envision a future machine learning algorithm that does not result in general discontent between the living and nonliving. Or, as the most pessimistic among those in the study of race have affirmed and reaffirmed, any use of technology has been shown to do little more than satisfy the premonitory dream of whiteness and its set of mythological valuations, including more efficient racial hierarchization. Machine learning, functioning as work and information for whiteness and its consumerist subsidiary, turns into the despoiler of individual and community life, the conquest of a world that finds well-being in its technological creations and the exploitation of energies both in terms of material resources and fields of human vitality. This will to power expresses itself in the technophile, the technocratic, as well as the technophobe that use technology and the idea of general progress as a cover to conceal the abundance of conscious/unconscious biases, implicit/explicit racisms, segregations, violences and other reductions of life chances that regulate our view of the human condition.

Machine learning, in its present conception, echoes an abundance of regulatory governance of the element of individual freedom, as read through an affirmative way of life, and a collective value distorted under our current and dynamic schemas of uncertainty about the future. Individual elements are denied independence from these logics of governance—a logic of enumeration—that demand adherence to the aspirations of external power. The resultant individual and collective self operates within the system under an expressed negation of life at the expense of a self-actualized future. A regulatory relation of circular causality, as such, cannot be established if the codes that comprise humans and machines merely rely on the experience of humans working on or working against a governed reality and the function of its authority. In this way, individual and collective life is reduced to one giant mathematical problem. To do so prohibits the crucial information of life from achieving its authentic purpose as an expression of the correlative existence of humans and algorithms and the value that they might inspire. The symbol of this aspiration is weakened into a mere technological or social exercise, whereby governance is allowed to step in at signs of impoverishment with our own self-actualization.

As Oscar H. Gandy, Jr., argues, "most public decisions these days are made on the basis of some analysis of data."[6] In *The Panoptic Sort: A Political Economy of Personal Information*, Gandy considers the role that statistical governance plays in "the reduction of life chances" under what he terms "a panoptic sort of data."[7] Gandy conceives of the "panoptic sort" as a type of data that extends beyond general surveillance. The panoptic sort is an all-seeing discriminatory apparatus that classifies individuals on the basis of their estimated economic or political value, and is continually optimized for the efficient transfer of value into data and information that dislocates and reassembles racialized bodies under the spatiotemporal objectives of institutions that maintain ownership over the circulation and classification of data.[8]

Gandy builds on Michel Foucault's 1975 book *Discipline and Punish* to address the intensification of discipline on the

6 Oscar H. Gandy, Jr., *Coming to Terms with Chance: Engaging Rational Discrimination and Cumulative Disadvantage* (Farnham: Ashgate, 2009), 4.

7 Oscar H. Gandy, Jr., *The Panoptic Sort: A Political Economy of Personal Information* (Boulder, CO: Westview Press, 1993), 15.

8 Adam Arvidsson, "On the 'Pre-History of the Panoptic Sort': Mobility in Market Research," *Surveillance & Society* 1, no. 4 (September 2002): 457.

racialized body.[9] Drawing on the concept of the panopticon, Gandy argues that the space of the prison is only one component of a wider distribution of power that leaks into the everyday activities of commerce. These spaces, which are circulated through statistical classifications, are formed around institutional and commercial operations and consumer value. For Gandy, statistical classifications reconfigure the universal position of surveillance, as they typically have a disproportionate effect on Black and racialized individuals.

What is most apparent, and perhaps just as troubling, is that while altering (or in the case of race, maintaining) orders of perception, governance can also form new modes of objectivities that operate under the weight of data analysis in support of racial difference.[10] Governance furthermore shifts accountability for these operations away from the production of a racialized substance of the social to the algorithm. Here, either the algorithm or the engineer is held accountable for the repetition of race and racism, while the substances of race that are a symptom of wider logics of enumeration remain untouchable and out of

9 In *Discipline and Punish* Foucault uses Jeremy Bentham's concept of the panopticon to illustrate the regulatory effects of surveillance and disciplinary power. The panopticon was designed by Bentham to maximize the surveillant perimetry of prison guards while minimizing direct control over detainees. To maximize guard visibility, the structure of the prison was reorganized in a circular layout with a control tower as the central focal point. In theory, a central tower would reduce the number of guards needed, as it provided complete and uninterrupted views of all prisoner cells along the inside circumference of the building. Individual prisoners were constantly subjected to bright lights originating from the tower, rendering their mundane activities visible, while also obscuring any view of the guards above. Spotlights prevented the prisoners from returning the surveillant gaze and distanced them from any corporeal forms of power, leaving only abstract visibility. In this configuration, the everyday activities of the surveilled individual are distributed as a mechanism of auto-regulatory control in the general population.

10 See Solon Barocas, "Data Mining and the Discourse on Discrimination" (paper presentation, Data Ethics Workshop, Conference of Knowledge Discovery and Data Mining, New York, August 24, 2014).

view. For instance, everyday transactions—from credit card transactions to browsing habits, customer reward programs, barcode scans, digital access points, biometric sampling, job applications, or drug testing—are the means by which Black people are targeted for exploitation, discrimination, redlining, criminality, and suspicion.[11] As Gandy suggests, any discourse on the biopolitical impact of data should therefore extend beyond the general sites of data aggregation. This is necessary to consider how the distribution of power aligns with the impositions of race and capitalism.

While racial discrimination in technology is not new, machine learning enables an unprecedented penetration of racial logics into the social milieu. These dynamics are clearly illustrated in the rapid adoption of machine learning algorithms for purposes of tracking, monitoring, and other forms of surveillance. Machine learning can mediate the consistency of racist logics by positioning the Black, brown, criminalized, and otherwise Othered body as always already sites of investigation. We see this most readily when machine learning algorithms produce racist and/or discriminatory outputs on widely recognized digital platforms, even when these algorithms are not designed with these intentions. In this way, while the algorithm itself does not comprehend, at least computationally, the complexities of race or racism, it can be perceived as racist when outputs simulate existing racialized human dynamics.

---

11 Seeta Peña Gangadharan defines data discrimination as the "processes of algorithmically driven decision-making and their connection to injustice and unfairness in society." Since algorithms are code that instruct computers to perform in certain capacities and data are discreet elements of our identity as representatives of the real, then it is safe to say that discrimination can be considered as a process by which data is continually instructed to drive unfairness into our cultural landscapes. In other words, what is articulated through data and machine learning is a mechanism of power concealed as objective decision-making. See Seeta Peña Gangadharan, with Virginia Eubanks and Solon Barocas, eds., *Data and Discrimination: Collected Essays* (Washington, DC: New America and Open Technology Institute, 2014), https://www.newamerica.org/oti/policy-papers/data-and-discrimination.

When power does reveal itself through the algorithmic, it remains obscure, frequently shifting accountability to the algorithm. In this sense, racial discrimination is often framed as algorithmic error or computational inefficiency that requires further capital investment in technological solutions. These mechanisms do not merely reinstate power. They are, as Sylvia Wynter argues, a bioepistemic relation that flows through systems of sociality while altering the very nature of racial perception.[12] With the notion of the bioepistemic, Wynter alerts us to the pervasiveness of epistemic forms of power and how the production of knowledge becomes the principal weapon by which difference is formed and reiterated. These knowledges are enacted through sciences that serve as proxies for objective forms of truth and natural law. Nonetheless, a key element of the bioepistemic is what Wynter describes as a *substance* of race. According to Wynter, the substance of race is a precondition for the staging of white European males as a basis from which all difference is derived. The substance, backed by science, gives the white European male a fictive sense of position at the apex of species, and consequently an implicit and explicit sense of power over the Other.[13]

Wynter speaks of science in more general terms as the weaponization of Eurocentric logics of racial difference, and machine learning reopens and reinstantiates these logics at an unprecedented rate. Here, a connection can be made between the notion of the bioepistemic and Foucault's formulations of biopower. Foucault warns against the preservation of empirical standards as a type of power or biopolitics that is enforced through the ordering of space, the prescription of behavior, and disciplinary actions against those who do not adhere to

12 Sylvia Wynter, "Human Being as Noun? Or *Being Human* as Praxis? Towards the Autopoietic Turn/Overturn: A Manifesto" (unpublished essay, 2007), 70. Available at https://www.scribd.com/document/329082323/Human-Being-as-Noun-Or-Being-Human-as-Praxis-Towards-the-Autopoietic-Turn-Overturn-A-Manifesto.

13 Wynter, 7.

the measures set forth.[14] By the second half of the eighteenth century, however, this power transformed into techniques that did not do away with discipline per se, but evolved into "nondisciplinary" instruments dissolved into everyday social practices. Here, Foucault points toward broader distributions of power that extend into civic operations, such as inspections, bookkeeping, reporting procedures, surveillance, and formal accumulations of data. Foucault argues that "after a seizure of power over the body in an individualizing mode, we have a second seizure of power that is not individualizing, but, if you like, massifying, that is directed not at man-as-body but as man-as-species."[15]

Foucault outlines a trinodal schematic that characterizes our reliance on the "truth-effects" that power produces.[16] First are the "rules of right that power implements to produce discourses of truth"—rule-governed and legitimated forms of power that do not have single centers, but instead traverse, characterize, and constitute the dynamic social body.[17] They are emboldened by the normalizing logics of species in accordance with classificatory operations that enact judgment, condemnation, and the forced performance of mundane tasks. A key characteristic of these truth-effects are illusions of transcendental embodiment, or imposition. What makes this type of power unique is the way in which the body becomes a mediator between ideological truth and subject formation. This is illustrated in the construction of race as a category through which phenotype can be understood in relation to the ideal subject to power. Illusions of ideal forms of subjectivity, according to Foucault, play a central role in the invention of the straight European white male as the ideal figure of *Man*—against which the Other is measured, sorted, and arranged.[18] The fictive Man represents an alteration in the

14 Michel Foucault, *"Society Must Be Defended": Lectures at the Collège de France, 1975–1976*, trans. David Macey (London: Penguin, 2008), 243.

15 Foucault, 243.

16 Foucault, 24.

17 Foucault, 24.

18 Foucault, 184.

"functional arrangements or systemic organizations" of the Other as the expressed truth of natural law.[19] On this grid, the Other is constituted by relations that are at once quantifiable and calculable—in terms of their deviation from normalized standards of behavior. In other words, for Foucault, order is that which brings the Other into confrontation with the law and the Enlightenment values of clarity and transparency. As he describes in *The Order of Things*, the Other is legitimated by exposure to values that link behaviors that are incongruent (or incompossible) with the approximate averaging of things, "in which fragments of a large number of possible orders glitter separately in the dimension, without law or geometry, of the *heteroclite* [...] in such a state things are 'laid,' 'placed,' 'arranged' in sites so very different from one another that it is impossible to find a place of residence for them, to define a *common locus* beneath them all."[20]

Foucault asks: "What type of power is it that is capable of producing discourses of power that have, in a society like ours, such powerful effects?"[21] To answer this question, Foucault urges us to look beyond the ownership of power and rather invest in the discovery of the real practices of power, as well as the applicable effects these powers have on the constitution of the subject: "In other words, rather than asking ourselves what the sovereign looks like from on high we are implored to discover how multiple bodies, forces, energies, matters, desires, thoughts, and so on are gradually, progressively, actually and materially constituted as subjects, or as the subject."[22] Foucault cautions against power as viewed as a phenomenon of mass or homogeneous domination, or as one that looks on high through a chain of events. He writes: "Power functions. Power is exercised through networks, and individuals do not simply circulate in those networks; they are in a position to both submit to and exercise this power. They are

19 Foucault, 7.

20 Michel Foucault, *The Order of Things: An Archaeology of the Human Sciences*, trans. Alan Sheridan (London: Routledge, 2007), xix.

21 Foucault, *"Society Must Be Defended,"* 24.

22 Foucault, 28.

never the inert or consenting targets of power; they are always its relays. In other words, power passes through individuals. It is not applied to them."[23]

The relation Foucault describes is a preexisting arrangement that stages data as a common locus that rests beneath the displacement of the racialized. Data, in this sense, are more than operative forms of value; they establish the values of difference as the subject and object of organization. They organize space into "interstitial blanks" that are constructed under the fictive relationship between difference and idealized models of species.[24] To a large extent, the distribution of these types of calculi are also preconditioned by the fictive substances of race. Foucault is critical of discourses on race that center displaced groups as the site of power, which he argues reinstates classifications of difference that portray racial distinction as contrary to the idealized model of Man. While attempts at intervening in racial discourse might produce some form of equitability, they also reestablish power as that which can regulate social behavior. The illusion sets forth the conditions for the defense of certain racialized bodies over others as biological threats to natural, and therefore societal, equilibrium. The principles of race serve as calculable principles of Otherness, exclusion, and segregation.[25] Here society directs its attention (and intentions) inward in search of entities that might articulate disharmony or disorder, while they simultaneously stage the conditions for incompatibility within wider ideals of harmony. Foucault argues that these logics are distinct from any concrete justification for their operations, which widens the gap between the logics of exception and any real sense of understanding of these mechanisms by the public.[26]

---

23 Foucault, 29.

24 Foucault, *The Order of Things*, xvii.

25 See Ezekiel Dixon-Román, "Toward a Hauntology on Data: On the Sociopolitical Forces of Data Assemblages," *Research in Education* 98, no. 1 (August 14, 2017): 44–58.

26 See Michel Foucault, *Security, Territory, Population: Lectures at the Collège de France, 1977–1978*, trans. Graham Burchell (New York: Palgrave Macmillan, 2009).

Why Fanon?

An immediate parallel can be drawn between Foucault's outline of biopower and Fanon's accounts of subject formation. While both Fanon and Foucault are concerned with the distribution of power, they diverge on the initial conditions from which the Other is constituted. Foucault presupposes a more general distribution of the means of power that brings the Other into view. Although Fanon does not mention Foucault explicitly, he is critical of discourse that prioritizes the means of subjection as universally embodied. Fanon places an unwavering emphasis on colonialism and the naturalization of race as the basis from which all sociality is understood. Here, the racialized subject as object of white culture does not enter into a universal relation with the distribution of power. The categorical is not merely the site of power's operation. It is the precise location in which racialized beings are excised from the category of human. Fanon is clear in *Black Skin, White Masks* that the logics of exclusion, as well as the epistemic operation of categorization—or what he describes as a "drama" of discovery—precede the Enlightenment principles of ideal subjectivity. This composition extends beyond the corporeal body and into the universal perception of Blackness, which is exposed by stereotype and emboldened by the distributed power of interpellation. It rearticulates the framing of life and death, put forward by Foucault, as that which exposes the racialized being to negating effects of history.

Although Fanon's comments on technology are limited, he does suggest that the fragmented body is not a passive vessel.[27] The collision between the fragmented body—as a racialized object—and the technical object articulates a relation of inequitable power distribution, as well as a site of contestation. For

27 Fanon comments on social diagnosis and the embodied effects of surveillance practices in a section of notes on a series of lectures titled "Le contrôle et la surveillance," translated in English as "Surveillance and Control." See Simone Browne, *Dark Matters: On the Surveillance of Blackness* (Durham, NC: Duke University Press, 2015).

Fanon, technology can be a means by which the colonized body can auto-generate new forms of radical subjectivity. For example, he references the CCTV (closed circuit television) camera as a localized form of surveillance on the everyday activities of colonized Algerians. Here, he preempts panoptic biopolitics by pointing toward the dissonances between the identification of social deviancy and the continual monitoring of public and worker activities as suspicious, and potentially criminal. The camera, he argues, serves as a unidirectional eye which sees all but can never be gazed back upon. Fanon shows that while the camera is installed and made visible by shop owners under the explicit premise of customer theft deterrence, it also organizes spaces of labor. Workers are monitored by an uninterrupted stream of their movements and actions within the shop. Fanon notes, however, that in response to the explicitness of managerial oversight, workers often use the means available to them to perform acts of resistance. This might materialize through acts that would disrupt the operations of power, such as taking unexpected sick leaves, displaying boredom, and being late or not showing up at all for work shifts.

Digital Epidermis

The abovementioned acts of resistance might appear passive, but, as Simone Browne argues, they can also be seen as intentional means of contestation. The result is what Browne calls "digital epidermalization," or methods by which power is exercised through the disembodiment of the Other under the gaze of surveillance and other technologies.[28] Browne demonstrates the fragility of the technological gaze that is enacted under the alienating logics of truth and categorical reasoning. Nonetheless, in doing so, Browne builds upon the dissonant relationship Black peoples have had historically with Anglo-centric technologies. As Browne argues, understanding this relation

28 Browne, 110.

is fundamental to any discourse on surveillance and the ethics of technology. This is particularly important considering the prevalence of discourse that centers the technical object as the subject of investigation without thorough (if any) insight into how technology shapes social space.

In *Dark Matters: On the Surveillance of Blackness*, Browne returns to Fanon as a point of departure to consider how anti-Black violence has helped shape technical knowledge and surveillance, while simultaneously rendering Black bodies nonexistent. In her introduction, Browne describes a correspondence she initiated with the CIA and the FBI concerning documents the US government held on Fanon produced during his travels to the United States to receive treatment for myeloid leukemia at National Institutes of Health Clinical Center in Bethesda, Maryland. Fanon was a patient there from October 10, 1961, until his death on December 6, 1961, at the age of thirty-six. In response to her Freedom of Information Act (FOIA) request, Browne received four declassified files from the FBI that included: a newspaper clipping about Fanon; a memo marked "secret"; a book review of David Caute's 1970 biography *Frantz Fanon*, filed under "extremist matters"; and other previously classified documents that mentioned Fanon's potential influence on Black Panther Party members and "amongst young Negroes." The FBI files—which Browne notes are heavily redacted—form traces of US government surveillance activities against Black radicalism, including details on Fanon's physical presence in the United States, as well as any intellectual and visual influence he may have had on Black youth. In his biography, Caute notes that the files profiled Fanon as being "no friend of the United States or of a free society."[29]

The detail of the FBI files contrasts the more obscure files released by the CIA under the same act. The CIA exercised Executive Order 13526 in response to Browne's FOIA request, which allows the agency to "neither confirm nor deny the existence or nonexistence of requested records."[30] The response states that

29 Browne, 1.

30 Browne, 1.

"the fact of the existence or nonexistence of requested records is currently and properly classified and is intelligence sources and methods information that is protected from disclosure."[31] It is here that Browne argues against public spaces that are shaped by and through whiteness, which are operative through acts of normalization, coding, and disciplinary measures. Data enacts a new form of power, through which prejudicial practices become sites where fear and terror are realized. While the Black body, as the subject of these operations, is restricted in terms of social, political, and economic access, it is also asked to willfully participate in its own regulation. A reflection on this regulatory exchange is crucial, considering what Browne calls her stalled "surveillance of the records of the FBI's surveillance," the "willful absenting of the record and the state's disavowal of the bureaucratic traces of Fanon," and how data takes shape as state power.[32] Notable are the points at which the existence of the corporeal body is rendered human only inasmuch as it maintains a compliance to technological governance. Here, to exist as a Black being in the public realm is to avoid punitive redaction from the public sphere, which Browne posits is part of a long history of attempts to regulate the classification and declassification of the Black body.[33] By connecting data to power and knowledge, researchers can uncover how data might replicate the immediacies of racism and discrimination. As Browne has shown, the logics of classification are enduring in their ability to stall the building of self-knowledge in the present while also regulating the existence of certain bodies, even after death. They also speak to the immediate shaping of public space.

Fanon is insistent that any discourse on the relation between the corporeal body and its environment is impotent without due consideration of how this relation regulates the humanness of Black beings. The colonial episteme presupposes life as being always already conditioned for social categorization. More so,

31 Browne, 1.
32 Browne, 3, 2.
33 Browne, 6.

it positions whiteness as the baseline of social and phenotypical measurement, as well as the fundamental principle against which one can be classified as human. In this sense, one might think of classification as an epistemic judgment over life and death. In contrast to early modern colonial occupation, Achille Mbembe argues that high technology tools, or the "spoils of modernity"—such as the gunboat, steam engine, submarine telegraph cables, railroad infrastructures, and so forth—have been weaponized throughout history to enact full unmitigated rights over segregation and the death of subjugated populations. These rights are distributed to institutions as well as citizens as exercises of power and sovereignty. He writes: "To exercise sovereignty is to exercise control over mortality and to define life as the deployment and manifestation of power."[34] This power enables the "generalized instrumentalization of human existence and the material destruction of human bodies and populations."[35]

While colonial operations are largely conceived of as operations of history, in Mbembe's view, what might be thought of as the distant past is indeed another stage of imperialism that involves the configuration of key technologies for colonial aims.[36]

---

34 Achille Mbembe, "Necropolitics," *Public Culture* 15, no. 1 (January 1, 2003): 12.

35 Mbembe, 14.

36 Anne McClintock makes a distinction between the logics of imperialism and the operation of colonialism, as the two are often conflated. McClintock explains, "Imperialism is not something that happened elsewhere—a disagreeable fact of history external to Western identity. Rather, imperialism and the invention of race were fundamental aspects of Western, industrial modernity." She goes to further describe the nuances of imperialism and colonialism: "European imperialism was, from the outset, a violent encounter with preexisting hierarchies of power that took shape not as the unfolding of its own inner destiny but as untidy, opportunistic interference with other regimes of power." Anne McClintock, *Imperial Leather: Race, Gender and Sexuality in the Colonial Contest* (New York: Routledge, 1995), 5, 6.

Imperialism and colonization are much more than merely systems of labor exchange. Imperialism is bound to the spaces of everyday life. It is useful to consider, then, that imperialism is an action that involves the extension of power by the acquisition and

Informed by Hannah Arendt's account of the seizing and delimiting of territories by Nazi regimes during World War II and segregation in apartheid South Africa, Mbembe elaborates on the function of colonial expansion as a historical consistency of segregation, mass discretizations of space, and human extermination. In this way, technologies enable the spatiotemporal repositioning of the Other into new sets of social, economic, and political relations. This, what Mbembe calls the "politics of exception," grants rights of life and protection to the most dominant, while sanctioning the death of those outside of the category of normality. The necropolitical process reinforces the boundaries of separation that produce hierarchies and distribute sovereignty as forms of life. Violence and subjection, in this sense, are articulations of the colonial imaginary, and are furthermore predicated on the blueprints of fragmentation, identity, and death as necropolitical consequences of imperialist ideas. Mbembe writes that this exercise "involves the setting of boundaries and internal frontiers epitomized by barracks and police stations; it is regulated by the language of pure force, immediate presence, and frequent and direct action; and it is premised on the principle of reciprocal exclusivity."[37]

We see throughout Mbembe's work how the historical process is linked to notions of technological advancement and terror, taking seriously the critique of actionable operations of reason as the state of exception with endless justifications for the "concatenation of multiple powers: disciplinary, biopolitical, and necropolitical."[38] The presence and consistency of these powers

discretization of territories. Colonialism, on the other hand, is the expression of imperialism in everyday life. Mbembe extends this framework of power into the unfolding of restrictions of movement to implement separations along the model of the discriminatory state, where "a patchwork of overlapping and incomplete rights to rule emerges, inextricably superimposed and tangled, in which different de facto juridical instances are geographically interwoven and plural allegiances, asymmetrical suzerainties, and enclaves abound." Mbembe, "Necropolitics," 31–32.

37 Mbembe, 26.

38 Mbembe, 29.

is totalizing, regularized, and endlessly shifting with regard to subjection. Mbembe expands on Foucault's concept of biopower to consider how the splintering of populations creates a further divide between the individual's ability to self-generate (or in terms of necropolitics, self-terminate) and the distribution of the colonial fantasy of sovereignty: "The objective of this process is twofold: to render any movement impossible and to implement separation along the model of the apartheid state. The occupied territories are therefore divided into a web of intricate internal borders and various isolated cells."[39]

Similar to Mbembe, in "Necrologies," Eugene Thacker extends the notion of exceptional life under the conditions of perpetual insecurity—a life that is "constantly rendered in its precariousness, a life that is always potentially under attack and therefore always an exceptional life."[40] Thacker builds upon Giorgio Agamben's concept of bare life to reconceptualize what we think of as biopolitics to emphasize the ordering of the body politic and its recurrent resurrections. Thacker draws a link between the biopolitical order and the medicalization of the human body. Within the necrological operation, Thacker argues, the proper functionality of the political body is to protect itself from the threat of dysfunctionality, and as such is continually attempting to preserve its own life, while discretizing the body into the parts that pose threats to the whole. Building on the continuity between biopolitics and necropolitics, Thacker suggests that the neoliberal notion of politics operates through the continued folding of what he calls the "whatever-life" into biological sovereignty into "forms of nonhuman life that are the agents of attack."[41]

In light of this, Fanon's work on pathological alienation can be adapted to rearticulate our present understandings of data, and in turn reimagine how machine learning might be

---

39 Mbembe, 28.

40 Eugene Thacker, "Necrologies; or, the Death of the Body Politic," in *Beyond Biopolitics*, ed. Patricia Ticineto Clough and Craig Willse (Durham, NC: Duke University Press, 2011), 158.

41 Thacker, 159.

viewed outside of representation. Fanon allows us to consider how any discourses today on machine learning (as well as race and machine learning) might be always already predicated on the exclusion of the Black body. We must consider whether any echoing of racism and racial hierarchy might be not a problem of mathematics, but a symptom of a larger exclusionary exercise that predates machine learning technology. The Black psyche might affirmatively auto-actualize within racialized systems. Fanon's profound solution in *Black Skin, White Masks* is that that the legitimacy of the racial imaginary can be disrupted through auto-actualization when the individual takes on a psychic refusal to representation. For Fanon, the individual, both Black and non-Black, who refuses to stay in past representation or consider the present as definitive is disalienated, or lodged free of oppressive representation. This idea has a considerable impact on contemporary thought concerning machine learning. Fanon's radical humanism asserts that the reclamation of individual freedom is found by prioritizing the future, whereas data prioritizes the past as a prediction of a future that exists only by mathematical certainty.

Fanon's disengagement with historical circumstance is crucial in considering the capacity to adopt a resistance to representation. It suggests a responsibility for the individual to "travel" through the dynamic creation of subjectivity to (re)imagine new representations and ways of living. This process, although not void of historical influence, is not only a necessary element of freedom but a politics in itself, particularly as data becomes an increasingly important factor in the social and political subjection of the individual. The psychic implications of this "certain uncertainty" must be examined in this contemporary space.[42] Machine learning's convergence with the Black psyche can then be understood as a relation of preestablished categorization by which practices of power, as well as Black psychic generation, might be examined more closely. Machine learning is a fairly young technology, which has increased in popularity in the last

42 Fanon, *Black Skin, White Masks*, 463.

few decades. It is through machine learning that we might gain new methods to liberate Black psychic generation from negating forms of power.

Fanon argues that possible resolutions to negation are found in invention—not in the form of the object such as gunpowder or the compass, but though accounts of one's own lived experiences. The conditions of these relations collide with the potential for self-actualization and the rebuilding of social spaces. In its present form, machine learning can in many cases paralyze processes of self-actualization. In this sense, mathematical rule sets and symbolic reasons take priority over the fostering of human conviviality. Furthermore, an unmitigated weight is placed on data through the valorization of difference, which necessitates the reinvention (or reimagining) of the anthropomorphic dilemma between the classification of nature and the embodiment of the logics of normalcy. Here, in order to understand how we can release data from the racializing effects of truth and order, we must first understand how data and machine learning have emerged as a proxy for individual and collective decision-making.

While these concerns appear to be recent installments in the historical trajectories of race and technologies—particularly given the exponential increase in machine learning algorithms and contemporary debates on data justice—they are underwritten by a much larger genealogy of racial sorting, as well as the distribution of bioepistemic modes of control. These milieus lead one into believing that reality is indeed connectionist, while simultaneously in displacement of the complexities of human and technological life. The directive is false, as is our dependence on its self-correction. It is, as argued by Rob Kitchin, only effective at "abstracting the world into categories, measures and other representational forms—numbers, characters, symbols, images, sounds, electromagnetic waves, bits—that constitute the building blocks from which information and knowledge are created."[43]

43 Rob Kitchin, *The Data Revolution: Big Data, Open Data, Data Infrastructures and Their Consequences* (London: SAGE, 2014).

Machine learning, while it involves a minor mode of existence that replicates the human condition via abstraction and categorization in its technological function, is an expression of that which has already been categorized. In this way, machine learning is not causal of racial categorization. Machine learning is the expression of a signal that alerts us to an individual and collective condition that finds its most prized value in the categorization of life in order to facilitate meaning through external affirmation.

It is useful to consider machine learning as more than a detrimental tool that can penetrate the very fabric of Black being. Any action required to mitigate these concerns is situated beyond an ethical or regulatory framework. What is needed is a methodological shift in the way we view data and machine learning as remedies to social problems. To understand machine learning as already imbued with the epistemic logics of racial exclusion is to recognize that the recognition of Blackness, as such, is always already incompatible with the algorithmic milieu. The minor and major modes of access between machine learning and the milieus of racially informed categorization must be explored, if only to build a knowledge upon which new approaches to the Black technical object can emerge as an aspirational potential from within the substances of race.

ACT I
On the Audacity of Black Aspiration

CHAPTER 1
As If: Critical Thoughts on Gaining Access to Black Aspiration

> We cannot represent ourselves. We can't be represented.
>
> —Stefano Harney and Fred Moten

In 2016, Joy Buolamwini, a researcher with the Civic Media group at the MIT Media Lab and founder of Code4Rights, developed Aspire Mirror. Aspire Mirror describes itself as a device that allows one to "see a reflection of [their] face based on what inspires [them] or what [they] hope to empathize with."[1] The project draws inspiration from the Thing from the Future, a card game that asks "players to collaboratively and competitively describe objects from a range of alternative futures."[2] Aspire Mirror draws additional influence from futuristic machines and speculative imaginaries found in popular science-fiction novels (for instance, the empathy box and mood organ in Philip K. Dick's *Do Androids Dream of Electric Sheep?*, stories of shape-shifting from the Ghanaian tale of the spider Anasi, and movies like *Transformers*). Buolamwini says she developed Aspire Mirror to induce empathies that can help facilitate the spread of compassion. Aspire Mirror also seeks to catalyze individual reflection based on a set of cultural values like humility, dedication, oneness with nature, harmony, faith, and self-actualization.[3] Ultimately for Buolamwini, these transformative futures are a "hall of possibilities" where individuals can explore self-determinant futures, "if only for a small period of time."[4]

Aspire Mirror relies on facial detection and tracking software to capture and interpret image data before transforming them into futuristic scenes or "paintings." During testing, Buolamwini encountered a problem. Aspire Mirror could not detect details of her presence due to her dark skin tones and facial features. In order to validate the device and generate an

1 "Idea," Aspire Mirror, http://www.aspiremirror.com/#idea.
2 "The Thing from the Future," Situation Lab, https://situationlab.org/project/the-thing-from-the-future/
3 "SciFi," Aspire Mirror, http://www.aspiremirror.com/#scifi.
4 "Design Process," Aspire Mirror, http://www.aspiremirror.com/#design-process.

alternative reality, Buolamwini had to first alter her existing appearance to make herself visible and gain access to aspirational futures. She accomplished this by wearing a white facial mask, with features that were more easily detected. For Buolamwini, this was no surprise. Buolamwini had encountered this limitation before when developing a computer vision system.

A Brief Note on Computer Vision and Machine Perception

Computer or machine vision is a broad term used to describe methods for processing and analyzing high-dimensional data acquired from the "real world" in order to produce symbolic or numerical outputs. The processed data can then be used to form decisions. Computer vision is powered by machine learning and artificial intelligence algorithms, and is commonly used for facial and image detection, modeling, and aesthetic judgment. It is derived from various bodies of research, including optics, theories of light, surface modeling, and auditory analysis. It is used to identify structures using internally represented models of objects previously known to the computer system. These models, usually geometric, find image features that match the model features with the right shape and position. An advantage of this technique is that the model encodes the object shape, thus allowing predictions of image data and lessening the chance of coincidental features being falsely recognized. The primary aim of computer vision is to interpret visual array to understand objects detected in the environment.

As Brian Potetz and Tai-Sing Lee write, computer vision models that infer 3D structures from 2D models are highly complex and underconstrained, requiring many assumptions to infer 3D shape from image points or environmental training data. Little is known about the merits of these assumptions in real scenarios. There is an awareness, however, that these assumptions must be simplified using probabilistic priors, which are based on real scenes. In fact, exact inference is only possible for a small subclass of potential problems in which case approximations must be used. Furthermore, Potetz and Lee note that developing human 3D surface interfaces (most often using Bayesian techniques) may uncover entirely new sources of information not immediately obvious from real physical models. They write: "Real scenes are affected by many regularities in the environment, such as the natural geometry of objects, the arrangements of objects in space, natural distributions of light and regularities in the position of the observer."[5]

While Potetz and Lee see these trends as exploitable in terms of algorithm design, few studies have considered the implication in terms of the model's own computational reasoning. Here, the model, despite its potential usefulness, maintains an operational perception of irregular and unobservable data to calculate the disparity between probabilistic priors and real scenarios. What would therefore be incomputable and unexplained is instead assigned a function and meaning in terms of the model's own logics of calculation. The algorithm assigns a logical potentiality within the perceptual distance between real scenarios' subsequent iterations of the learning process. This potentiality exists even if it is unknown how they might be realized. According to Christopher W. Tyler, these objects consist physically of aggregates of particles that cohere together to form the object under investigation. Tyler describes the process as such: "Although the objects may be stable and invariant in the scene before us, the cues that convey the presence of the objects to the eyes are much less stable. They may change in luminance or color, they may be disrupted by reflections or highlights or occlusion by intervening objects. The various cues carrying the information about the physical object structure, such as edge structure, binocular disparity, color, shading, texture, and motion vector fields, typically carry information that is inconsistent."[6]

These algorithmically organized particle structures emphasize an overriding problematic in machine learning research. In

5 Brian Potetz and Tai-Sing Lee, "Scene Statistics and 3D Surface Perception," in *Computer Vision: From Surfaces to 3D Objects*, ed. Christopher W. Tyler (Boca Raton, FL: Chapman and Hall, 2011), 2.

6 Christopher W. Tyler, "Introduction: The Role of Midlevel Surface Representation in 3D Object Encoding," in *Computer Vision*, vii.

order to accomplish what is thought to be visual coherence, it is functionally necessary to reduce any inherent levels of inconsistency or instability, as a human eye would. Here, coherence depends heavily on the mitigation of "occlusions" or blockages that may disrupt a clear view of the world around us. In other words, to perceive the world as such, the algorithm must functionally simulate the complexity of, say, a human's optical system by means of reduction and simplification.

Structuring perception through machine architectures is problematic when considering the racialized object. To bring forth a visually coherent future in machine learning, or an incoherent and speculative one for that matter (using Buolamwini's example), one must first reduce the hall of possibilities to a set of preexisting conditions. This is inherent in the function of machine perception. While machine perception has been defined in a number of ways (for instance, logics, self-awareness, reasoning, planning, and problem solving), attempts at goal achievement are considered distinct from relative indeterminate human experience. In terms of operational aims, machine perception functions by stabilizing and simulating environmental phenomenon in order to grasp the fundamental basis of human and other sentient-based actions, mimicking behavior and managing tasks thought to exceed human capabilities. This is seen, most readily, in robotics inspired by the living, which are both aesthetically and functionally created to solve complex problems thought to be out of reach of most human. Nonetheless, the basis of simulation here characterizes the living as an emanation of preexisting conditions, reducing the operation of individuation and the differences amongst the living to no more than an assemblage of contradictions that are negated and subsumed into a higher, more homogenous unity of existence.

Beyond Reality and the Problem of Locating Known Objects

Many coders, like Buolamwini, use pre-written coding libraries to perform common computer vision tasks. In the case of Aspire Mirror, this code includes libraries such as Beyond Reality Face for face detection and tracking, Vibrant.js for color extraction, jQuery for animation, and Chrome Remote Desktop as a panel control. Beyond Reality Face, for instance, is a cross platform and real-time face tracking software development kit (SDK) that marks faces in images and webcam streams. It is a proprietary and downloadable SDK comprised of a collection of resources, such as data, pre-written code, subroutines, classes, values, and type specifications. Beyond Reality Face analyzes faces using a morphable shape with sixty-eight individual feature points that indicate the eyes, nose, mouth, face, and other details found on the human head. Once the feature points are detected, Beyond Reality Face estimates the 3D position, rotation, and scale of the head, even while the head is in motion. It then uses the tracking data to overlay virtual objects onto images of the detected face. While the complete list of algorithms used in the SDK are proprietary to Tastenkunst, the company that developed Beyond Reality Face, it is known that the SDK relies on an optimized frames per second (FPS) algorithm and an active shape model (ASM) algorithm.

ASMs are algorithms that generate statistical models of the shape of objects in order to conform the suggested shape to the point distribution model. They are a type of model-based vision algorithm established as an approach to recognize and locate known objects in the presence of noise, clutter, and occlusion. ASM developers argue that while traditional model-based vision algorithms sacrifice model specificity in order to accommodate variability, ASMs instead fit the data in ways that are consistent with the training set, allowing developers to locate partially occluded objects in noisy, cluttered images within specific contexts determined by the training data.[7]

Regardless, common image detection libraries are often trained on normalized spectrums of data that are prone to false negatives without proper light conditions. In other words, they

7 See Tim F. Cootes et al., "Active Shape Models: Their Training and Application," *Computer Vision and Image Understanding* 61, no. 1 (January 1995): 38.

are trained on image data that includes primarily white subjects. The white phenotype then becomes the preexisting condition and the prototypical assemblage from which all future human characteristics are measured. As result, the darker the skin tones or the more variant a person's phenotypical features are from the average white subject, the less likely the algorithm is to recognize an individual's presence.

Toward the Black Technical Object

While Buolamwini notes that her experience with Aspire Mirror is telling of the functional limitations of computer vision and machine perception, she argues that the problematic is largely one of representation. For Buolamwini, a failure to recognize Black faces as coherent human objects demonstrates "a lack of diversity in the [data] training set [that] leads to an inability to easily characterize faces that do not fit the normal face derived from the training set."[8] At the same time she reminds us that "whoever codes the system embeds her views. Limited views create limited systems." This "coded gaze," as she calls it, is a potential catalyst for political action and technical intervention. "Let's code with a more expansive gaze," she writes. Buolamwini foregrounds an unwitting link between Black pathology and the technical object, what—in its convergence—we might call the "Black technical object." The link finds precedence in Fanon's thesis on the operations of race and its impact on the psyche of the racialized. As a practicing psychiatrist, Fanon put forward a line of propositions that extend beyond the corporeal implications of racism into the realm of the psychic and collective individual. For Fanon, the colonial project is an organic system that is demarcated by a series of relations between these individuals and violent forces of racism. It is essential to note that within

8 Joy Buolamwini, "InCoding—In the Beginning Was the Coded Gaze," Medium, May 16, 2016, https://medium.com/mit-media-lab/incoding-in-the-beginning-4e2a5c51a45d.

these relations, the individual's psychic fragmentation is made apparent to the individual at the moment of their encounter with racializing phenomena. Returning to Buolamwini's case, we can think in terms of the recognition of incompatibility between her own sense of self as a coder and the machine's functional perception of her as an undetectable, and therefore nonexistent, object. Within Fanon's schema, Buolamwini's undetectability exaggerates the dissonance between her own self-determination (she just wants to be seen) and the external forces (Aspire Mirror) that restrict this desire based on a predetermined rule set. The possibility of self-determination is opened only under the condition that the individual alters the self, which in this case is a transformation of the flesh.

Fanon's schema departs from the point of nonexistence to discuss the objectification of the individual through this process, which includes their own psychic fragmentation. Consequently, Fanon explains, the individual embodies a phobic image of the self that is infused with paranoia, delirium, and self-doubt. The phobic image of the racialized (or what he calls the "photogenic object") is embedded in the psychic orientation of the West.[9] Within these imaginary systems, self-doubt becomes the guiding principle by which the racialized person views themselves as well as the world around them. These affective relations are felt and acted on, effectively replacing what is actually seen with a fictive belief about how one should perceive the photogenic object. While the white psychic structure experiences the same phobia, according to Fanon, it is instead articulated as a social incompatibility or threat, where the racialized become visible as individual beings only inasmuch as their possibility for existence aligns with preexisting concepts of racial hierarchy.

If we consider the photogenic object in contemporary spaces of algorithmic culture, it is apparent that the Black technical object is always already preconditioned by an affective logic of race that functions on the psychic level of experience. The possibility of an affirmative engagement between the Black

9 Fanon, *Black Skin, White Masks*, 151.

technical object and the algorithm, as a technical object, is then limited by the necessity to reconcile the psychic potential of the racialized individual with that of a predetermined technical structure. Although the immediacy of computation's lack of diversity—in terms of institutional value and algorithmic function—cannot be understated, a call to make Black technical objects compatible with computer vision algorithms risks the further reduction of the lived potentiality of Black individuals and collectives. While the achievement of Buolamwini's aims might widen the scope of machine perception, not to mention the participation of excluded bodies in techno-social ecologies, the proposed solution reinforces the presupposition that coherence and detectability are necessary components of techno-human relations. The drive toward uniform coherence works to circumvent any substantial consideration of the algorithm's reliance on historical category—namely, which features represent the categories of human, gender, race, sexuality, and so on. Coders like Buolamwini speak directly to the problem of erasure, more specifically the erasure of being, yet the act folds seamlessly into a desire for representation. "Challenged to rethink, insurgent Black intellectuals and/or artists are looking at new ways to write and talk about race and representation, working to transform the image," as bell hooks writes.[10] hooks reminds us, however, that, "there is a direct and abiding connection between the maintenance of white supremacist patriarchy in this society and the institutionalization via mass media of specific images, representations of race, of blackness that support and maintain the oppression, exploitation, and overall domination of all people. [...] For it is only as one imagines 'woman' in the abstract, when woman becomes fiction or fantasy, can race not be seen as significant."[11]

Still, the institutionalization of representation is devoid of the dynamisms of Black life: "For some time now the critical

10 bell hooks, *Black Looks: Race and Representation* (Boston: South End Press, 1992), 2.

11 hooks, 2, 124.

challenge for black folks has been to expand the discussion of race and representation beyond debates about good and bad imagery. Often what is thought to be good is merely a reaction against representations created by white people that were blatantly stereotypical. Currently, however, we are bombarded by black folks creating and marketing similar stereotypical images. It is not an issue of 'us' and 'them.' The issue is really one of standpoint."[12]

In terms of computation, following hooks, we are immediately drawn to the problematic of Black representation and algorithmic calculation. By regressing complex environmental data into a generalized pattern, the lived and multivalent specificities of Black lives are represented "as if" the visual matrix maintains no connections to historical category, stereotype, or moral imaginary. It further highlights a commitment within research to organize certain life in accordance with the rules of occlusion, which already necessitates a presupposition of object relation (in terms of the Black face, that which is brought into view only inasmuch as it can make sense of the world in relation to the white body).

The consequences for the Black technical object are immense. An undetectable Black technical object is, in this instance, equivalent to a machinic nonexistence, or the formation of an empirical reality that—as Wynter articulates in her critique of Western humanism—is a projection of a racialized substance that is metaphysically sustained by the Aristotelian embodiment of normalcy.[13] Put another way, research in computation is an adaptation of the fictive and compulsive ordering of human attributes into a single coherent image of species, or what Wynter has described as "being secularly human."[14] Wynter draws on Judith Butler's critique of gender, in which Butler argues that Otherness emerges when we enact the nouns "man" and "woman" as abiding substances. Butler states that

12 hooks, 4.
13 Wynter, "Human Being as Noun?"
14 Wynter, 4.

these substances are produced by the fictive and compulsive ordering of attributes into a coherent gender sequence. If these coherences are nothing more than "contingently created through the regulation of attributes," Butler posits that any ontology of substance is itself an "artificial effect."[15] Wynter's adoption of this point of reference extends the artificiality of regulated attributes into the substances of class, sexual orientation, and race. Her claim is prompted by the creation of what she describes as eugenic/dysgenic selection.[16] The conditions for racial sorting and priority were already set forth in the establishment of data analysis and statistical correlation as viable tools for social inquiry. In many ways, the limits of calculation cannot be understood outside of the connection between racial sorting, social welfare, and quantification.

We can think of this relation as illustrative of what Ian Hacking describes as "enduring ways of thinking" that have impacted philosophical understandings of mathematics.[17] For instance, Hacking uses A. C. Crombie's text "Philosophical Presuppositions and Shifting Interpretations of Galileo" as an example of the reliance on "(a) the simple postulation and deduction in the mathematical sciences, (b) experimental exploration, (c) hypothetical construction of models of analogy, (d) ordering of variety by comparison and taxonomy, (e) statistical analysis of regularities of populations, and (f) historical derivation of genetic development."[18] He highlights these paradigms to note an important shift in social frameworks that increasingly rely upon reasoning to establish scientific method. This matrix of knowledge production, as Hacking argues, has given rise to the false idea that information on population phenomena can be accurately evaluated through symbolic sampling, as opposed to the exhaustive census work that had been attempted prior.

15 Butler, quoted in Wynter, 8.

16 Wynter, 8.

17 Ian Hacking, *The Taming of Chance* (Cambridge: Cambridge University Press, 1990), 6.

18 Hacking, 6.

These techniques were intensified by the work of Lambert Adolphe Jacques Quetelet and André-Michel Guerry. Having observed the characteristics of human life and dynamic natural phenomena, Quetelet and Guerry argue that a number of equivalences can be found in their behavioral characteristics. This is not to say that the rising of the sun is equivalent to a human rising out of bed, but that certain human behaviors can be predicted with the same level of certainty as the rising or falling of the sun.[19] As result, human behavior was thought to be bound to a causality, much like the universal constants of astronomical phenomena. As with the stars, all deviations from human constants of behavior are deduced as perturbations of naturalized events that, once settled back into an equilibrium, can be returned to previously normalized patterns. In other words, the hypothesis aligns contingent and chaotic phenomena with a statistical order. The data, however, are actionable in their capacity to infer future social behavior based on historical and present observation. From these technologies, prototypical behavior emerges as a mode of population control, mediated through what Johann Carl Friedrich Gauss termed the *homme type*, or the "average man."[20]

Quetelet's primary emphasis is on the codification of phenotypical phenomena, particularly race, inasmuch as race extends beyond the biological and into moral substance. Hacking writes:

> Where before one thought of a people in terms of its culture or its geography or its language or its rulers or its religion, Quetelet introduced a new objective measurable conception of a people. A race would be characterized by its measurements of physical and moral qualities, summed up in the average man of that race. This is half of the beginnings of eugenics, the other half being the reflection that one can introduce social policies that will either preserve or alter the average qualities of a race. In short, the average man led to both a new kind of information about populations and a new conception of how to control them.[21]

19 Hacking, 105–6.
20 Hacking, 105.
21 Hacking, 107.

In Hacking's analysis, Quetelet uses basic statistical probabilities to construct a matrix of phenotypical characteristics. Interestingly, he has little interest in deriving numerical averages of phenotypical factors such as height and weight (this would be the equivalent of preempting an individual's rate of childbirth at 2.2 children or that an individual might marry 1.6 times). The primary function of the matrix is to identify a causal relationship between an array of unique, yet arbitrary, properties, such as human chest size to vegetable harvest or moral attributes. Using the logics of astronomical discovery, Quetelet hypothesizes that any departures from normalized patterns of hereditary development are disruptions in the natural trajectory of evolution. More so, he believes these differences would (or could) be naturally deselected. Alternatively, any conformity to normalized attributes would be naturally selected for both survival and superiority among its species. As Hacking writes: "Quetelet had made mean stature, eye-color, artistic faculty and disease into real quantities. Once he had done that [...] deviation from the means was just natural deviation, deviation made by nature, and that could not be conceived of as errors."[22]

In terms of machine learning, the site of exclusion in Aspire Mirror is also the precise moment where the Black technical object is interpellated, fragmented, and organized around the white visual metaphor. The coherence of racialized attributes, or the fictive substance of race, links the dynamic instrumentalization of coherence found in computation to the "discursive negation of co-humaneness."[23] The negation immediately enacts a preexistent distance between the perception of oneself as a self-determinant being and a more formative system of abstract value. In other words, Black life as such, or Black being as self-constituted, is at a continued distance from the illusionary prototypical image. What is one to do?

To merely include a representational object in a computational milieu that has already positioned the white object as the

---

22 Hacking, 113.

23 Wynter, "Human Being as Noun?," 4.

prototypical characteristic catalyzes disruption superficially. The white object remains whole, while the object of difference is seen as alienated, fragmented, and lacking in comparison. From this perspective, the position of the Black technical object is lodged in a recurrent dialectic, where it attempts to valorize or recapture Black life from within the confines of normalized logics while simultaneously desiring to disrupt its hold. Can the Black technical object be conceptualized as outside of the dialectic between human and machine? Is there such a thing, borrowing from Fred Moten, as an aspirational Black life that can gain a right of refusal to representation? As such, would a universal computational gaze limit the self-determination of those who have little or no desire for inclusion in machine perception? Alternatively, as facial recognition becomes an increasingly important lens through which we understand the world, how can Black technical objects generate new possibilities outside of phenotypical calculation, prototypical correlation, and the generalization of category? How might we create a more affirmative view of the relation between the Black technical object and technology?

Perceptive Unities

It is apparent that the immediacy of these concerns mirrors the intensity at which algorithms populate the public domain. Recent debates surrounding data, artificial intelligence, and machine learning as a techno-human practice, particularly in terms of the discriminatory powers of classification and further reductions in the life chances of marginalized populations, have raised questions concerning the logics of machine perception and its impact on self-actualization. Most significant is a disparity between the act of existing/existence—particularly as it relates to differential human states of being (categories of race, gender, sexuality, and so on)—and the paradigms of epistemological operation. The result is no less real than the metaphysical gestures algorithms seek to appropriate, such as conflict resolution, discovery, compatibility, and invention. Any solution to the complexity of the relation between Black individuals and algorithms is only exacerbated by an overdependence on mathematics as truth, and that which already exists in computation as fact—particularly in circumstances such as police surveillance or managerial oversight, where face detection reaches beyond the exploration of aspirations toward the operative reduction of life chances.

Trials conducted in 2018 by the London Metropolitan Police Service (the Met) enact similar logics of perception. The Met invited holiday shoppers in Central London to take part in a trial of a facial-detection system used to identify suspects wanted by the police. To test the system, the Met mapped existing police image data onto the software. Once trained, the algorithm interacted with cameras at a specific location. The camera scanned faces in the crowd, feeding the image data captured back into the algorithm. The algorithm then compared the captured image data and correlated it to image representations of suspects flagged by the police. Data representations of the crowd were kept in the police database for weeks.[24] The Met's trial was part of a larger investigation into the use of facial detection software to identify criminal suspects. However, these attempts have been met with scrutiny in terms of privacy (officers are instructed to wear plain clothes) and the robustness of the system. An investigation by campaign group Big Brother Watch indicates an alarming number of false positives. For example, in the Met's trial at Notting Hill Carnival in 2016 and 2017, the system incorrectly flagged 102 people as potential suspects. All led to no arrests. This is in addition to data acquired from the South Wales Police that shows 2,451 false positives out of 2,685 so-called matches.[25]

Facial detection operations are not limited to blind trials. They are employed to solve a range of economic, social, or

---

24 "Central London in Facial Recognition Trial," *BBC News*, December 16, 2018, https://www.bbc.com/news/uk-england-london-46584184.

25 Chris Fox, "Face Recogntion Police Tools 'Staggeringly Inacurate,'" *BBC News*, May 15, 2018, https://www.bbc.com/news/technology-44089161.

political objectives which rely on a greater coherence of individuals and the environment for operational efficiency. For instance, an online retailer may desire to increase sales by identifying which of its present or future customers are likely to purchase certain goods or not; or when, if ever, they are likely to make a present or future purchase. A bank may want to gain a deeper understanding of its client base to determine which customers are most likely to default on a loan or a credit card bill. Once correlated, these customers can then be classified by risk and given either higher or lower interest rates. A judicial committee concerned with recidivism may want to make parole decisions based on the probability a convicted person will become a repeat offender. Or a police agency may attempt to optimize the distribution of police cruisers based on the probability a neighborhood will yield lower crime rates based on police presence. Here, the notion of objectivity is once again challenged, bringing forth what Ezekiel Dixon-Román terms "algo-ritmo," or the disciplining of the flesh. Dixon-Román writes:

> Regardless of the degree of human subjectivity behind the code for this algorithm, it is the case that the performative act of this algorithm can be a powerful force in shaping and disciplining the flesh. In this example, algo-ritmo quite explicitly disciplines the flesh and designates humanity into full humans and nonhumans. […] As an immanent act beyond human intervention, algo-ritmo is a performative force that may do more than simply reify "difference." With the ubiquity of algorithms in society, algo-ritmo has the capacity to reconfigure the boundaries of "difference" as well as further magnify the sedimentation of "difference."[26]

Dixon-Román points to the act of classification as not only the reification of difference, but also the reconfiguration of the ontologies of the human in relation to technology. In these instances, the process of computer vision begins at the presupposition by

26 Ezekiel J. Dixon-Román, "Algo-Ritmo: More-Than-Human Performative Acts and the Racializing Assemblages of Algorithmic Architectures," *Cultural Studies ↔ Critical Methodologies* 16, no. 5 (June 26, 2016): 8.

the human perceiver that the machine has already attained a state of objectification independent of the variability of human experience and knowledge. Problematics are compounded when technological research collides with public trust/distrust in data and automated systems that organize the complex realities of human relations. While the technologies illuminate sentient and computational concerns, they also unearth techno-human phenomena that might otherwise remain hidden or obscured from vision. In this way, machine learning is as much relational as it is operative in the shaping of perception "as if" they are direct simulations of sentient intelligences—as illustrated in desires for more "natural" or sentient-like behaviors in speech generation/recognition, robotics, and artificial intelligences.

Machine learning then becomes a thought act that privileges the view that individuals are created on the basis of coherence and categorical division. In this way, the relation between human reason and artificiality emerges as an immediate contradiction of perceptual domains. While one—the human—is based on a multivalent array of lived realities, the other is founded in a nominalism devoid of dynamism. This astonishing circumvention of indeterminacy naturalizes and objectifies the variant ways in which human beings live their lives to a degree that any mode of coexistence becomes no more than a transcendental presence brought forth by a single epistemological point of view. The operation of individuation is furthermore relegated to a series of representations among a falsely unified species. The consequence is the "freezing" of dynamic life into a homogenous milieu, and the immediate suppression of the lived conditions that catalyze certain individual transformations and subsequently inform future iterations of being.

If we are, in Maurice Merleau-Ponty's view, relational beings, then life "has a social atmosphere just as much as it has a flavor of mortality."[27] Yet, by bringing forth the fictive substance of race into the realm of the artificial, machine learning

27 Maurice Merleau-Ponty, *Phenomenology of Perception*, trans. Donald A. Landes (Abingdon: Routledge, 2003), 425.

functions as a series of "rudimental navigations," using Johanna Seibt's description, that operate under the logics of rule-based procedures, as if they are human.[28] If machine perception is to mimic anything other than preexisting hylomorphic association, then we must take into account that the machine is not human. The artificial cannot comprehend the full scope of life and human existence. It is through the rule-based procedure, the substance, that the machine maintains both its dependence on limits of metaphysics and the objectification of life itself. These demands supplement what Seibt argues is a "gradual increase of the regulatory dependence up to normativity."[29] Here, multivalent human perception is not enlisted to widen our understanding of ourselves or the world around us, but is instead of value to research only inasmuch it can assist in the demonstration of what machines can or cannot do.[30] The current ontological position, Seibt argues, can be described more precisely as tendencies to place value on the knowing-how of knowledge production as a process dependent upon the formulation of description in conversation with normative modes of performance.

> From a philosophical viewpoint it is a category mistake to assume that we can interact with anything—whether robot or human—as if it were a person. "Person" is not a descriptive predicate designating a feature, but a declarative-ascriptive predicate designing a response-dependent condition like "red" or "C#," with the distinctive difference that the response to the item in question is not perceptual but normative—it is undertaking a commitment to certain actions and omissions, in accordance with rights and obligations. […] When we call an entity a person we thereby, in the performance of that utterance, take on certain absolute normative commitments—in fact, to call something a person is

28 Johanna Seibt, "Varieties of the 'As If': Five Ways to Simulate an Action," in *Sociable Robots and the Future of Social Relations: Proceedings of Robo-Philosophy 2014*, ed. Johanna Seibt, Raul Hakli, and Marco Nørskov (Amsterdam: IOS Press, 2014), 100.

29 Johanna Seibt, "How to Naturalize Sensory Consciousness and Intentionality within a Process Monism with Normative Gradient: A Reading of Sellars," in *Sellars and His Legacy*, ed. James R. O'Shea (Oxford: Oxford University Press, 2016), 188.

30 See Seibt, "Varieties of the 'As If,'" 97–105.

> to do nothing else but to make these commitments. […] You cannot say "It is as if I hereby promise you…" nor "I promise you somewhat…" Similarly, if we treat some x as a person we are committed to taking x as a person, which means that we interact with x as a person.[31]

The role of normativity undertakes new significances in ethical debates surrounding the applications of artificial intelligence systems. At risk is the continued (re)production of the categorical other and the foreclosure of the lived conditions through which specific humans might determine if a machine is friend, foe, colleague, or neighbor. Luciana Parisi has shown that the "actualities that select, evaluate, transform, and produce data" expose the internal inconsistencies of rational-based systems, particularly functional mechanisms that advocate for the naturalization of reason.[32] Her argument calls into question the fundamental logic of the rule-based procedure and offers new opportunities to reassess the question of what actualities count as processes of artificial reason. Without a wider scope, debates on these matters remain incomplete in their characterization of algorithmic prejudices and social discriminations. Attempts at reconciling this arguably unsettled debate rely on a commitment to sufficiently characterize the constitution of a more affirmative process of machinic existence that can gain a totality in relation to artificial modes of perception. This proposal asks us to consider what is overlooked in machine perception, and in doing so dislodges both the ontological and functional process of machine perception from its roots in substantialist metaphysics. Machine perception here demands a new reflexive position that can generate alternative levels of operation.

---

31 Seibt, 100.

32 Luciana Parisi, *Contagious Architecture: Computation, Aesthetics, and Space* (Cambridge, MA: MIT Press, 2013), ix.

Affirmative Psychic Genesis

A revision of this field of perception demands a return to metaphysics and a reading of the genesis of being from the perspective of multivalent modes of reality. Although artificial systems are based on classification and descriptions, with tendencies of revision and prediction, there are other modes of logic that ontology has underarticulated. Whereas Siebt proposes a technological solution that can account for languages that do not contain terms for "human mental states, agentive goals, or social relations," Stefano Harney and Fred Moten suggest that we imagine an ontological relation that prioritizes the psychic generation of the Black technical object. Within this framework, we are asked to give thought to the Black technical object that does not "want to be correct" or "corrected":

> Consider the following statement: "There's nothing wrong with Blackness": What if this were the primitive axiom of a new Black studies underived from the psycho-politico-pathology of populations and its corollary theorization of the state or of state racism; an axiom derived, as all such axioms are, from the "runaway tongues" and eloquent vulgarities encrypted in works and days that turn out to be of the native or the slave only insofar as the fugitive is misrecognized, and in bare lives that turn out to be bare only insofar as no attention is paid to them, only insofar as such lives persist under the sign and weight of a closed question?[33]

Here, we can imagine an object that develops an indifference to description or any other form of artificial representation. It would maintain its lived experience as an internal transformation, both within and in excess of artificial perception. The act of psychological transformation here challenges the state of homogeneity by engaging in a transformative politics of affirmative self-belonging. Harney and Moten develop their understanding of the transformative subject by suggesting that the individual

33 Stefano Harney and Fred Moten, *The Undercommons: Fugitive Planning & Black Study* (Wivenhoe: Minor Compositions, 2013), 47–48.

viewed as not belonging is represented as a type of cultural entropy within dominant systems of power. However, this "hidden" individual is only excess insofar as it operates from the viewpoint of lack or decay. Harney and Moten argue that within the experience of social contact, what bell hooks might call a *communion*, the entropic individual exceeds the barriers of social relations to enter an alternative space of becoming—made possible by a reimagining of selfhood.[34] In other words, allowability for the unusable, uncommon, and thus incomputable individual potentializes the social space toward new ways of relating.

If, as Moten argues in "The Case of Blackness," "the cultural and political discourse on black pathology has been so pervasive that it could be said to constitute the background against which all representations of blacks, blackness, or (the color) black take place,"[35] then how might the pathological dulling of Black life inform new readings of computer vision? What would it take to resist notions of universal correlations, and instead value the dissonance that emerges in relation with analytics? By prioritizing cohesion, algorithmic processes erode the potential for human difference and self-actualization. Coherence gives rise to what Gilles Deleuze and Félix Guattari call the "demands of singularity," or an insistence on the union of diverse entities into a single group, form, body, or relation.[36] In *Nomadology: The War Machine*, Deleuze and Guattari remind us that "the model in question is one of becoming and heterogeneity, as opposed to the stable, the eternal, the identical, the constant."[37] The Black technical object converges with the artificial in an assemblage of mutable and multivalent experiences. Here, both the Black technical object and technical object inform each iteration of themselves in a self-governing system of feedback.

---

34 bell hooks, *Communion: The Female Search for Love (Love Song to the Nation)* (New York: William Morrow and Company, 2016), xviii.

35 Fred Moten, "The Case of Blackness," *Criticism* 50, no. 2 (Spring 2008): 177.

36 Gilles Deleuze and Félix Guattari, *Nomadology: The War Machine*, trans. Brian Massumi (New York: Semiotext(e), 1986), 18.

37 Deleuze and Guattari, 31.

While entities within the assemblage might be perceived as incompatible, new conditions for self-actualization emerge at each moment of contact. In other words, the Black sense of self is formed, informed, and reformed at the moment of dissonance between self-perception and any externally constructed view of Black life. These tensions are the conditions from which new iterations of the self are generated, exceeding the reductions of representation and visibility. They form a recurrent system of feedback that enacts what Foucault has called a technology of the self, which "permit[s] individuals to effect by their own means or with the help of others a certain number of operations on their own bodies and souls, thoughts, conduct, and way of being, so as to transform themselves in order to attain a certain state of happiness, purity, wisdom, perfection or immortality."[38] By prioritizing difference, entropy (or the loss of computational coherence as such) can be conceptualized as the condition through which transformation is made possible. In this way, entropy is revealed to be a system inclusive of contingency, instability, multivalent modes of perceptions, indeterminacies, and iterations of self-actualization.

If we accept the present matrix of computer vision as a normalizing logic, then perhaps we should turn away from our dependencies on the artificial to activate the internal halls of human potential. The concerns that arise from within our parasitic relation with the technical object are no less immediate, as artificial intelligences such as computer vision articulate a wider logic of reductionism and Black exclusion. What we experience today as algorithmic prejudice is the materialization of an overriding logic of correlation and hierarchy hidden under the illusion of objectivity. Meanwhile, the fictive substances of race and racialization work to disrupt self-actualization by reenforcing the false assumption of coherence. In the drive toward coherence,

38 Michel Foucault, "Technologies of the Self," in *Technologies of the Self: A Seminar with Michel Foucault*, ed. Luther H. Martin, Huck Gutman, and Patrick H. Hutton (Amherst: University of Massachusetts Press, 1988), 18.

computer vision is set in place as if it is human and the guardian of judgment. In operation, it is assigned the role of interpellator, assigning value (in terms of visibility) to the individual only inasmuch as they can be measured against a universalizing concept of being. In the collision between Blackness and the artificial, this operation can materialize, even unwittingly, as an incoherence, which places undue weight on the perception of oneself and environment. While this perception speaks to the immediacy of racism and racialization, an opportunity emerges to shift the pathological perspective from one of entropy and lack to a more affirmative process of psychic generation. Here, the development of a machinic existence would, at its origin, substitute the view of computational duress to one of Black totality, always already in the process of transformation. The resultant incompatibility could then be seen as an act that, while bringing forth preexisting substances of racialization, can make use of this duress to catalyze future affirmative iterations of the self. Ultimately, what is prioritized within this psychic encounter is a compassion for the self as already coherent at the encounter of artificial misrecognition—a self that is continually taking shape, as Blackness has always done, in its exploration of infinite halls of possibility.

CHAPTER 2
Sociogeny and the Recapitulation of Racialized Being

> I hope by analyzing it to destroy it.
>
> —Frantz Fanon

Fanon and the Problem of Ontogenesis

In *Black Skin, White Masks*, Fanon stages his move toward authentic disalienation following Sigmund Freud's insistence on prioritizing the question of ontogenesis. In psychoanalysis, ontogenesis is a theory of development of an individual viewed from the perspective of occurrences or events throughout their life. According to Freud, the events of an individual's life constitute the prehistory of their early childhood, which has become unconscious to the individual, resulting in neurosis.[1] In the psychoanalytic space, once the influence of the ontogenetic events of an individual's life and its possibilities have been exhausted, ontogenesis should be replaced by phylogenic analysis, or the study of an individual's whole family (denoting tribe, race, or one's evolutionary relatedness) to better illuminate one's prehistory. An individual's prehistory is mainly discovered by analysis of their memories as well as the memories of their relatives. Phylogenesis is useful in biology and psychoanalysis as it helps the individual understand their evolutionary relatedness to their closest population. Freud largely believes that although humans are not at the center of the universe, the human species does suffer from psychic delusions of power over other species. Freud's ultimate aim is to elaborate on the evolutionary causes of mental syndromes in the megalomaniac patient, who is in a crisis of self-expression.[2]

For Freud, the role of psychoanalysis—as a science psychic exploration—is to intervene as mediator between the megalomanic individual and their external environment. Freud finds particular value in Charles Darwin's theories of evolution,

---

1 Sigmund Freud, *The "Wolfman" and Other Cases*, trans. Louise Adey Huish (London: Penguin, 2002), 215.

2 Freud, 215.

specifically the biological principle that "ontogeny recapitulates phylogeny."[3] The biological notion of recapitulation provided a method by which Freud could gain a more complete understanding of the human psychic development. The principle stipulates that an individual's development occurs through a series of stages that are then duplicated, or recapitulated. The recapitulation process contributing to the historical evolution of the whole of the human species. For instance, consider the outdated thought that embryonic gill slits in humans can be seen as a recapitulation of early humans or so-called lower life forms. Gills are thought to indicate the biological truth of individual evolutionary development, as well as the lineage of human evolution.

It has been argued that Freud's application of these recapitulations in psychoanalytic treatment is largely self-serving. M. Kathryn Armistead has written extensively on Freud's misuse of recapitulation to establish misguided links between the anthropological primitive or "savage" and individual patient neurosis.[4] Armistead points to the first documentation of the nineteenth-century law ("ontogeny recapitulates phylogeny") drafted by Ernst von Haeckel. Armistead notes that although Haeckel and others documented apparent visible similarities in comparative organisms, these theories had been adopted in psychology, sociology, and education without verifiable proof of these discoveries among humans. According to Armistead, Freud's adoption of the principle, and application to psychoanalysis, rests on a set of assumptions later adopted by the social sciences to validate anthropological findings.

Although some psychoanalytic studies, particularly those by Richard E. Nisbett and Lee Ross are critical of psychoanalytic theories that reify racial biases and in turn substantiate

---

3 See G. R. Searle, *Eugenics and Politics in Britain: 1900–1914* (Leyden: Noordhoff International Publishing, 1976).

4 M. Kathryn Armistead, "A Critical Examination of Freud's Scientific Premise that Ontogeny Recapitulates Phylogeny in Totem and Taboo," *Didache* 9, no. 2 (January 2010): 3–4.

scientifically reinforced stereotypes, psychoanalysis has been for the most part silent on the effects of economic and political disparities in certain communities.[5] The exclusion constitutes what Ralph Cintron describes as "discourses of measurement,"[6] or what Laura Doyle shows to be a constructed hierarchy in science and nature. She writes: "while seeming to privilege the organic world biological theories from phrenology to evolution insidiously turn that domain against itself, using it to give evidence of supremacy of mind and of the race and sex most skilled in the arts of the mind (including, in a circularly self-authorizing move, the scientific arts)."[7]

Fanon is adamant that Freud's self-assumptions are symptomatic of a larger ontological misdirection that prioritizes the development of white Europeans as the basis from which all other human populations are analyzed. Recapitulation perpetuates beliefs in a natural order of species development. Fanon contends that this ontological structure provided justification for the European colonization, and in many cases "extermination," of peoples of the Global South.[8] He maintains that ontology as such is dismissive of the actual phylogenetic process, which for racialized being is not a natural progression. For the racialized, the sequence of events involved in the evolutionary development of colonized people is not nature. The developmental structures of the colonized are contaminated by a succession of events that are characterized by death, violence, and exploitation at the hands of white European colonialists. On this ontological failure, Fanon argues that "society, unlike biochemical processes,

5 See Stanley O. Gaines Jr., "Perspectives of Du Bois and Fanon on the Psychology of Oppression," in *Fanon: A Critical Reader*, ed. Lewis R. Gordon, T. Denean Sharpley-Whiting, and Renée T. White (New York: Wiley-Blackwell, 1996), 24–33. See also Richard E. Nisbett and Lee Ross, *Human Inference: Strategies and Shortcomings of Social Judgment* (Englewood Cliffs, NJ: Prentice-Hall, 1980).

6 Ralph Cintron, *Angels' Town:* Chero *Ways, Gang Life, and Rhetorics of the Everyday* (Boston: Beacon Press, 1999).

7 Laura Doyle, *Bordering on the Body: The Racial Matrix of Modern Fiction and Culture* (New York: Oxford University Press, 1994), 7.

8 See Fanon, *Black Skin, White Masks*, 18.

cannot escape human influences." Man, he states, "is what brings society into being."[9] Then he offers a solution.

In the opening chapter of *Black Skin, White Masks*, "The Negro and Language," Fanon elaborates on the concept of sociogeny. Sociogeny, he writes, is the product of the historical realities of anti-Blackness, whose outcome is the epidermalization, or internalization, of the racial being's psychic inferiority.[10] Sociogeny is a "sociodiagnostic" phenomenon, whereby as a result of social factors such as colonialism and racism, the racial being internalizes how they are perceived by the dominant other—a feat that he argues is embodied in the internal/external perception of the racial other as cultural object.[11] Sociogeny is that which conditions psychic individual and collective individuation, committing the racial being, as Fanon argues, to recurrent cycles of neurosis and social objectification. As a practicing psychiatrist, psychic genesis is of central importance to Fanon. Although he situates the internal/external psychic relation within the mind/body dialectic (indicating his commitment to contracting revisions of representation as opposed to subverting it in its entirety), he does prioritize the recurrent (and, I would argue, parasitic) relation between external formations of being and internal psychic development. The specificities of his assessment, or how he situates the distorted formation of the racial being's cognitive relation to the colonial imaginary, is in direct confrontation with the exteriority of racial logics as reinforced by the ontological adoption of being as such. It is within this discursive "flaw" that Fanon argues for a revision of theories of being founded on the misrecognition of the sociogenetic conditions from which they are derived. Sociogeny is situated within a mind/body dialectic, indicating Fanon's commitment to challenging notions of representation.

---

9 Fanon, 4.

10 Fanon proposes the concept of epidermalization to illustrate the objectification of the Black colonial subject. Epidermalization is a way of thinking about the racialized body as object, or as it experiences "being through others." Fanon, 109.

11 Fanon, 109.

For Fanon, the spaces between the ontological concept of being and the recurrent cycles of racial hierarchy are disruptive to the process of self-actualization, whereby one instead desires to "win admittance into the white world [… in] a constant effort to run away from his own individuality, to annihilate his own presence."[12] This is also where one can turn toward the new law of the algorithmic to defend their claim for recognition, where they might one day find that the "psychological minus-value, this feeling of insignificance and its corollary, the impossibility of reaching the light" has not vanished as they once expected.[13] An enormous task confronts those who turn to the machine for mutual recognition. Something remarkable, if not entirely fantastical, happens whereby the Black practitioner is faced with a duplicate of delirium unchanged: one from the human condition of racial commitment and the other from the agonizing conviction of the white world to build technology in its own image.

To the dismay of some technocrats, machine learning is not a bilateral process. A familiar neurosis places us in direct confrontation with the immediate recognition of the psychosocial realities of our own aspirations to be something other than being-through-others, be they human or machine. A machinic sociogenesis seeks to take into account a more complete understanding of racialized occurrences involved in the evolutionary development of machine learning, whereby algorithmic racism as witnessed today becomes symptomatic of a larger interruption in individual and collective maturation. A prognosis not only rests in the hands of those willing to unearth and expel the toxicity of systemic racism in the algorithm, but also those individuals who are willing to dislodge any imposed neuroticism.

12 Fanon, 42–43.
13 Fanon, 44.

Black Mirrors and the Problematic "I"

> I resolved, since it was impossible for me to get away from an inborn complex, to assert myself as a BLACK MAN. Since the other hesitated to recognize me, there remained only one solution: to make myself known.
>
> —Frantz Fanon

Fanon reminds us that "we cannot afford to forget that [...] the neurotic's fate remains in his own hands."[14] For Fanon, this fate is not one of self-determination. He writes: "For the black man there is only one destiny. And it is white."[15] Fanon's efforts to reconcile the colonized individual and racialization is materialized in psychosocial studies of oppression and resistance. He contends that the internal self must confront the particulars of external colonial environments; the individual must end their own internal colonization. Otherwise, they will continue to be consumed by repressed anger, resentment, and aggression brought on by colonial operations. Here, following Jean-Paul Sartre, the individual can authenticate a being that is no longer condemned by race, and is therefore freed and disalienated from a position of shame.

Fanon's radical self-affirmation is, in this sense, largely a reaction to existentialist accounts of transcendental subjectivity and the systematic totality of knowledge. The emergence of structuralism in the late 1940s and early 1950s provided Fanon with an opening to insert the specificity of the colonial subject into consideration of subjective genesis, namely by drawing upon the structuralist psychosocial framing of the "I" developed by Jacques Lacan. While it is important to note that Fanon does not explicitly describe his patients' childhoods, he does pathologize the presupposition of neurotic illness within individual and collective development. In the Lacanian notion of the mirror phase, the child's perception of themselves is experienced as a

14 Fanon, 4.

15 Fanon, 13.

fragmentation of the self in reflection to the wider world around them, which can be perceived but not directly accessed. As Lacan describes: the comparison allows the child to "anticipate in a mirage the maturation of his power [which] is given to him only as *Gestalt*, that is to say, in an exterior in which this form is certainly more constituent than constituted."[16] It is tempting to pull away from the "I" for fear of a return to structuralism and therefore an anthropocentric view of the world; however, the emergence of structuralism provided Fanon with a discourse that could better account for the "fact of Blackness"—which, Fanon argues, is the fundamental realization that the Black body both is generated and self-generates outside of the discursive concept of human.

Drawing from Lacan, Fanon situates the Black being's desire for recognition as a problem of the individual's cognitive structuring of the world. When the object of desire is expressed as a craving for external recognition, the individual faces an internal crisis that drives them toward a misrecognition of the internal self. As result, the individual becomes self-alienated in the act of recognition, which Fanon argues is expressed through self-doubt and frequent comparison of one's self to the other. Fanon is pointing toward the genesis of the psyche and the organization of collective self-doubt, which he evidences in the ontological denial of Black life. The denial of Black life gives rise to Fanon's critique of Lacan's ontological misdirection. Lacan posits that the imaginary structuring of one's conception of reality results in a "permanence of I" that alienates and prefigures the subject into a fictional image of themselves.[17] He uses Claude Lévi-Strauss's notion of the effectiveness of symbols to discuss the gravity of the specular image as the threshold of the

16 Jacques Lacan, *The Seminar of Jacques Lacan*, ed. Jacques-Alain Miller, trans. Alan Sheridan, vol. 6, *The Four Fundamental Concepts of Psychoanalysis* (New York: W. W. Norton, 1981), 503.

17 Jacques Lacan, *Écrits*, trans. Bruce Fink (New York: W. W. Norton, 2006), 80.

visible world in our daily experiences.[18] The resultant psychosocial experience confines the Black subject into a permanent fragmentation that is always already lacking in a desire for the transcendent, which under colonialism acts as the category of the white gaze. In other words, the perception of the self and the world is crucial for understanding the fragmentation of the psyche, not to mention perceptive boundaries of the environment. This imaginary also produces fragmentations and inconsistencies that heighten sensitivity to psychic conflict. Still, Fanon's use of Lacan's lectures raises questions about the significance of symbolic function. For instance, it prompts us to think about the relationship between Lacan's "veiled faces" and W. E. B. Du Bois's metaphor of "the veil," introduced in the first chapter of *The Souls of Black Folk*, "Of Our Spiritual Strivings." For Du Bois, the veil has spiritual as well as philosophical associations. Along with the "double consciousness," the veil is one of two metaphors Du Bois uses to illustrate the spiritual strivings of former slaves and the complexity of their lives in post-emancipation America.[19] Under segregation, Blacks are afforded "no true self-consciousness" in the sense that their awareness of themselves, or how they see themselves, is measured only through the eyes of the other.[20] There is a sense of "always looking," Du Bois posits, onto a world that looks back in "amused contempt and pity."[21]

Du Bois outlines racialized experience from two perspectives. First is the act of racial subjection that "divides the soul" into a being that is mocked by the history of slavery while simultaneously aspiring to be one's "truer self" as something other than a rejection of the white world.[22] Let's pause on the idea of the "truer self"—not from the perspective of a spiritual striving toward a new essentialism, but thinking in terms of the

18 See Markos Zafiropoulos, *Lacan and Lévi-Strauss or the Return to Freud (1951–1957)*, trans. John Hollard (London: Karnac, 2010).

19 W. E. B. Du Bois, *The Souls of Black Folk* (Chicago: A. C. McClurg & Co., 1903), 8.

20 Du Bois, 8.

21 Du Bois, 8.

22 Du Bois, 8–9.

value created by the internal recognition of one's capacity for self-actualization while working through the duress of racial subjection. From this perspective, racial duress is the catalyst for affirmative psychic maturation, if only to build a clear psychic awareness of the necessity to prioritize one's aspirations in the midst of internal and external conflict. Du Bois finds power in this contradiction: "Throughout history, the powers of single black men flash here and there like falling stars, and die sometimes before the world has rightly gauged their brightness. [...] The black man's turning hither and thither in hesitant and doubtful striving has often made his very strength to lose effectiveness, to seem like absence of power, like weakness. And yet it is not weakness,—it is the contradiction of double aims."[23]

Although few would situate Du Bois directly in line with psychosocial study, his ontology shares a relation with concepts put forward by Fanon. Fanon asks that we realign our ontological assumptions toward a less reductive account of Black lived experience. Fanon relies on visual representations and technology to craft a reconstruction of the being that is situated from within an accumulation of knowledge that results in direct participation in events that effect Black development.

However, for both Fanon and Du Bois, the individual is the dominant agent in the process of self-actualization. Their theories are as self-determinant as they are radically optimistic, in that they suggest that the pathway toward an authentic negotiation with external duress can be built through an awareness of one's capacity for self-development.[24] The tensions between the internalization of race and conditioning oneself to reconstruct the internal relation deepens Fanon's formulation of space and time—as well as the fragmented self—as inhabiting neither position. His stance suggests that subject formation instead oscillates between these positions, resisting what Homi K. Bhabha describes as a "tradition of representation that conceives of identity as the satisfaction of a totalizing, plenitudinous object

23 Du Bois, 3.
24 Fanon, *Black Skin, White Masks*, 232.

of vision."[25] In other words, beneath Fanon's schema of visibility is a process of ongoing self-making built upon the efficacy of Black consciousness. But what becomes of the symbolic "I" in awareness of its sociogenetic conditions?

The "I" enacted under these terms overdetermines life and death as a performance that is inscribed under the ownership of knowledge and selection. Any coherence here misstates the conditions for the development of the "I," which is embedded—under the Lacanian framework—as an a priori enclosure of pathological lack. Here, the psyche is no more than an untenanted vessel of colonial occupation to be shaped and reformed at the will of any "I" that represents itself as dominant in the construction of learning. The enactment of Black life within the sociogenetic context is dulled under the totalizing conditions of race. Fanon contextualizes the problem of sociogeny from within the specificity of colonialism. However, I question the consistency of sociogeny. In other words, if we are to return to the ontological problem of origin, how might this coincide with other forms of episteme?

Fanon's Revision of the Master-Slave Dialectic

Self-consciousness is, in this way, a dynamic, intentional, presupposing interconnectedness, reinforced and confirmed by the struggle for recognition. The master-slave relation, as the most primitive of intersubjectivity, emerges as a complex array of transcendental forms—a phenomenon that Georg Wilhelm Friedrich Hegel characterizes as a "battle for recognition" in *The Phenomenology of Mind*.[26] Although the human becomes conscious of the self through desire, the subject also asserts itself as the center of the relation. The human is, thereby, abstracted

25 Homi K. Bhabha, *The Location of Culture* (London: Routledge, 2004), 46.

26 Georg Wilhelm Friedrich Hegel, *The Phenomenology of Mind*, trans. J. B. Baillie (London: Digiread, 2009).

by the transcendental forces of life, maintaining a distance from the phenomena of the world and their own self-actualization. In this sense, to gain recognition one must overcome any sense of internal superiority over one's environment. Here, freedom remains a construct of the rational, a notion that enunciates the binary between mind/body, human/nature, and so on.[27] In Hegel's reading of slavery, the slave lacks the courage to resist tyranny in an open contest, and can achieve freedom only through discipline, work ethic, and Christian forgiveness. The impact of the dialectic is what Sarah E. Chinn describes as "the abstraction of the body into a collection of measurable functions [which] renders it legible as a sign of something else, not itself: patterns, qualities, trends, predictable processes."[28]

The Phenomenology of Mind by was an important work for Fanon, particularly in its intellectual description of the origin and dynamics of oppression. In the text, Hegel argues that Man becomes conscious of himself through a desire for recognition by the Other. Those who are recognized but do not recognize others become the master, and those who recognize without being recognized become the slaves. Not only does the master gain recognition through the slave, he also reduces the slave to an object of his own will. According to Hegel, the slave, on the other hand, gains a sense of self through labor, at the bequest of the master. Yet, the relationship is not a convivial story for the master, who is alienated by his distance from labor and, as result, loses his means of transforming the world and himself. The slave—as object—reflects the master's perception of his own humanity, not to mention the master's inability to attain self-actualization. On the other hand, through his struggle for

27 See Joseph Young, "A Reversal of the Racialization of History in Hegel's Master/Slave Dialectic (Douglass's 'Heroic Slave' and Melville's 'Benito Cereno')," in *Race and the Foundations of Knowledge: Cultural Amnesia in the Academy,* ed. Joseph Young and Jana Evans Braziel (Chicago: University of Illinois Press, 2006), 94–113.

28 Sarah E. Chinn, *Technology and the Logic of American Racism: A Cultural History of the Body as Evidence* (London: Bloomsbury Publishing, 2000), 5.

recognition, and detachment from the fear of physical death, the slave gains the means to resist the master and transform his own humanity.

Hegel, as with Freud, ignores the specificities of oppression in European colonization and slavery. Hegel also assumes that the master held a superior position in the master-slave relation. In the chapter "The Negro and Hegel" of *Black Skin, White Masks*, Fanon—greatly influenced by Sartre—adopts the Hegelian master-slave dialectic as a paradigm to examine the contemporary relationship between white (masters) and Black (slaves). However, Fanon rejects the abstract relation set forth by Hegel's universal approach and rearticulates the dialectic to account for the immediate conditions of colonialism.

Fanon is critical of Hegel's positioning of the slave as cowardly and subservient. Fanon instead favors accounts of the slave as self-affirming, even if alienated under the conditions of the master-slave dialectic. Fanon argues that the experience of Black people in relation to whites excludes any ontological neutrality. In other words, Black people are always already objectified under the gaze of the "white other." As such, they do not maintain the same subjective position to reciprocate the gesture of looking upon and abstracting the white body. The lived experience of the Black individual is not a typical Hegelian dialectic between Self and Other, but instead a concealment of very specific inequitable relations constructed in the operations of fabricated histories, narratives, and patterns of oppressive colonial ideologies. As Fanon articulates, "For not only must the black man be black; he must be black in relation to the white man."[29] Within the Fanonian dialectic, the Black individual "outlaws any ontological explanation" of the ontological refusal of Black realities.[30] The Black subject is self-objectifying in confrontation with their psychic dislocation from the category of human. Fanon describes the perils of his own self-ascription as such: "I was responsible at the same time for my body, for my race, and my ancestors.

29 Fanon, *Black Skin, White Masks*, 82–83.

30 Fanon, 83.

I subjected myself to an objective examination, I discovered my blackness, my ethnic characteristics [...] completely disoriented, unable to be abroad with the other, the white man, who unmercifully imprisoned me, I took myself far off from my own presence, far indeed, and made myself an object."[31]

Fanon's clinical observations of shattered psyches, as in his work in Algeria, emphasizes his self-identification with the dimensions of Europe's colonial assaults. Although Fanon draws on Hegel to engineer a new mode of recognition, he remains optimistic that the Black subject's dignity and self-worth can be gained through a social relation that instead prioritized reciprocal recognitions of difference that resist "savage struggle[s]" for recognition.[32] Otherwise, as Fanon proposes, one becomes steeped in wretched servitude and objecthood.

According to Diana Fuss, the Black person's desire for the white Other coincides with the white "protean imaginary."[33] Fuss argues that white protean imaginary has come to legislate the category of human. It is also the measure against which the human species gains full, or even partial, cultural significance. Furthermore, Fuss notes that although identity has a temporal and spatial structure, the idea of a universal mirror image is challenged by the key signifiers of racial difference. In this way, the Black individual resides at the periphery of the normalized category of Man—a process that is enacted through systems of enumeration under colonial operation.

When social space takes shape within the larger cultural context of colonial expansion and imperialism, historical identification is supplanted by the construction of normalized colonial histories. While colonialism works to suspend spatial and temporal mobility (in both its physical and psychosocial sense), it also operates by regulating the boundaries of cultural knowledge. The implication is an immediate and devastating exclusion

31 Fanon, 84–85.
32 Fanon, 170.
33 Diana Fuss, "Interior Colonies: Frantz Fanon and the Politics of Identification," *Diacritics* 24, no. 2/3 (1994): 21.

from the freedoms of movement afforded to white privilege under colonial rule. This ensures that Blacks are sealed into a "crushing objecthood" under normalized logics of relation.[34] Here, the oppressor enacts procedures by which the racialized see themselves as neither "I" nor "not-I."[35] At the same time, Fuss reminds us that while the imaginary may signify white protean development in the non-Black other, the Black individual is fragmented and objectified. Fuss writes: "Black may be a protean imaginary other for white, but for itself is a stationary 'object'; objecthood substituting for true alterity, blocks the migration through the Other necessary for subjectivity to take place. [...] The Black man (contra Lacan) begins and ends violently fragmented."[36] Fuss draws her argument from Fanon's account of his own internal position as nonbeing:

> Sealed into that crushing objecthood, I turned beseechingly to others. Their attention was a liberation, running over my body suddenly abraded into nonbeing, endowing me once more with an agility that I had thought lost, and by taking me out of the world, restoring me to it. But just as I reached the other side, I stumbled, and the movements, the attitudes, the glances of the other fixed me there, in the sense in which a chemical solution is fixed by a dye. I was indignant; I demanded an explanation. Nothing happened. I burst apart. Now the fragments have been put together again by another self.[37]

Fanon's accounts are further complicated by what he considers to be an abandonment of any consideration of Black existence within ontologies of being. He writes, "Ontology—once it is finally admitted as leaving existence by the wayside—does not permit us to understand the being of the black man."[38] As Fanon observes, the psychoanalytic interpretation of the Black psyche emphasizes prolonged flaws in Black being that are reinforced by the instrumentalization of normality. A primary premise

34 Fuss, 22.
35 Fuss, 21.
36 Fuss, 21–22.
37 Fanon, *Black Skin, White Masks*, 89.
38 Fanon, 82.

of psychoanalytic study is that human behavior is an internal problematic for individuals. In other words, behavior is located "within" us. Nonetheless, for Fanon, psychoanalysis has failed to trace the processes by which our current epistemic systems have come to represent how humans are perceived, and perceive themselves, within the deeply contextualized Eurocentric paradigm. The primary concern of Fanon's thesis is the discontinuity of the racialized human as an integrated yet excluded category of species; and the establishment of the non-Black other as the prototype of the ideal human, and subsequently a perception of humanity as if it were a purely natural organism in continuity with organic life.[39] Still, the resultant crisis of this deductive "truth" starts at the specificity of Black psychic fragmentation—the psychosocial alienation of the Black subject and the self-justification of the disciplines of science in the solidification of the human sorting, in particular. There are key similarities between Du Bois's and Fanon's prioritization of self-actualization as a form of alienated and fragmentation of the Black psyche.

Both Fanon and Du Bois's accounts of the phenomenological experience of anti-Black racism are essential components to their critique of philosophies of the being. In Du Bois's analysis, he illustrates the concept of double consciousness—a recurring theme throughout his work—particularly in his attempts to reposition the freed American slave from within the Hegelian problematic. For Du Bois, the formerly enslaved person is confronted with a duality of existence comprised of the historicity of violence and subjection and the potential of newly emerging modes of exchange. Du Bois sees value in the latter as a potential path of self-actualization. He, however, shifts the focus of perception away from the white gaze (a schematic of oppression that he relegates to the present, as informed by consistency of racialization) and onto the potentiality of self-formation, which he articulates within the terms of prototypicality. He advocates for a Black sense of self modeled after what he considered to

39 See Sylvia Wynter, *No Humans Involved* (Hudson, NY: Publication Studio, 2015).

be exemplary forms of labor and representation—the teacher, preacher, bourgeois intellectual, and laborer.[40] In doing so, he suggests that Blacks can make new claims for recognition through a revision of the normative premises of social value. In doing so, the Black can free themselves from the continual hold of slavery and white oppression. Du Bois alerts us to the gift of "second-sight" that the Black individual attains when they are confronted with the question of their position within the human category. He focuses on the ways in which the code of race is integrated into cultural beliefs, as well as the psychological basis of racialized individuals. This gift of second sight, Du Bois proclaims, "only lets [the Black American] see himself through the revelation of the other world," which he describes as a "peculiar sensation" or "double-consciousness" that manifests itself as a subjective generation irreversibly bonded with the view of one's self through the eyes of others. Accordingly, the "measuring" of one's soul results in an inherent "twoness" comprised of "two souls, two thoughts, two unreconcilable strivings; two warring ideals in one dark body, whose dogged strength alone keeps it from being torn asunder."

Robert Gooding-Williams posits that the Du Boisean second sight can be distinguished from the more general Hegelian framework in that it instead characterizes the Black individual's capacity to see what is not visible with an extrasensory perception of history as well as the present conditions of anti-Black racism. With second sight, the Black individual has a heightened view of what Du Bois calls the "other world," which is a view of themselves through the gaze of whiteness.[41] In this sense, Gooding-Williams argues that Black peoples are "gifted" with the white's perception of them—which itself is deeply distorted by racism.[42] Nonetheless, Gooding-Williams argues that this gift also constitutes an internal striving or desire to be the white

40 Du Bois, *The Souls of Black Folk*, 8.

41 Du Bois, 9.

42 Robert Gooding-Williams, *In the Shadow of Du Bois: Afro-Modern Political Thought in America* (Cambridge, MA: Harvard University Press, 2011), 1.

Other. It fosters the Black's own self-contempt and presumed inferiority. In other words, for Du Bois and Fanon the Black psyche articulates more than just a stationary living organism that can be reduced to biological classification, but stages the problem of Black psychic development as a problem of the sociogenetic organization of social exchange.

Visual Perception and Fanon's Adoption of Gestalt

This position, as an organization of perception, resists generality, and is instead articulated from within the specificities of racialized experience. While Fanon stages this psychic problematic as one of alienation, he is careful to prioritize the reciprocity of desire as a meta-organization of spatiotemporal structures. These desires of Self and Other do not give way to self-determination, but are always already in conversation with universal prototypes of being. Fanon's sociodiagnostic scheme, if thought through as a nonlinear process of psychic generation, as opposed to a negation of self and environment, can provide new concepts of relation between the contingent capacities of machine learning algorithms and the actualization of more affirmative Black modes of perception.

Some underlying advantages of Fanon's work are the associations he makes between the spatiotemporal organization of racialized beings and the epistemic operations of colonialism. Kara Keeling argues that the ontological predicament of the Black man is situated at the epistemological limits—or what she calls the "coordinate"—of the colonial imaginary. Here, the colonial world shackles life to the past as a way of rationalizing colonial existence while also reinstating Black inferiority.[43] As Fanon posits, racism is not a constant of the human spirit. It is expressed as an obedience to a "flawless logic" that "draws its

43 Kara Keeling, "'In the Interval': Frantz Fanon and the 'Problems' of Visual Representation," *Qui Parle* 13, no. 2 (December 1, 2003): 97.

substance from the exploitation of other peoples."[44] He describes the disposition of racism as a latency of psycho-affective and economic relations that result in the cultural abandonment of the racialized individual, a futile alienation, as well as the individual's psychological "craving" for "History" and "Truth." Fanon writes, "Discovering the futility of his alienation, his progressive deprivation, the inferiorized individual, after this phase of deculturation, of extraneousness, comes back to his original position."[45]

Fanon refers to this position as the liquidation of the racialized individual's system of reference. The colonizer, as Fanon argues, instead takes ultimate authority over the spatiotemporal position of the colonized. The colonizer manages the grid of relation by imposing "new ways of seeing and, in particular, a pejorative judgement with respect to [the oppressor's] original forms of existing."[46] This often emerges through scientific "Truth" (which Fanon establishes as the "original position") and the day-to-day activities of the racialized. As result, the racialized can only recognize their occupancy in the colonial world through the invention of an ideal subject—the Anglo-European secular man. For Fanon, spatiotemporal organization serves an additional function. By claiming ultimate authority over the management of colonial space, the colonizer is able to construct an image of himself as the ideal secular being, one that is sanctioned by illusions of divine will. These illusions were supported by the distribution of scientific knowledge, which—in terms of the racialized—is enacted through eugenics-based operations.

According to Fanon, the epistemic relation subtends quantification, and thus perception, of the Black body as a negated form of species. Fanon defines this construction of the European model of Man as a "drama" of scientific discovery that stages the racialized body as a problem of nature, or what he describes as

44 Frantz Fanon, *Toward the African Revolution: Political Essays*, trans. Haakon Chevalier (New York: Grove Press, 1994), 40–41.

45 Fanon, 27.

46 Fanon, 2.

the divine abandonment of Black existence. For the racialized "God," he exclaims, "is not on [their] side."[47] As result, that science enacts new modes of perception in racialized environments organized around the provenance as substitute for the colonial imaginary. These modes extend into the construction of secular norms, as well as pathological alienation. Still, Fanon is careful to suggest that the complexities of the Black object relation are too often supplanted by concepts of racial transcendence and racial "consciousness." In this way, perception suspends the Black body as already situated at the intersection of phenotype and the call of the interpellator.[48]

Fanon uses visual metaphors to illustrate how his own experience as a Black man is organized around the fantasy of the white child. This is articulated clearly in his oft-cited passage: "Mama, see the Negro! I'm frightened!"[49] The passage is widely viewed as the foundation of Fanon's schema of enumeration and theory of race as a discourse that continually reifies the categorization of the Black body. While evidence points to Fanon's often uncomfortable and pessimistic images of Blackness in the text,

---

47 Fanon, 38.

48 Judith Butler describes Louis Althusser's act of interpellation as a way of "staging the call" for the individual to turn around and respond to oneself and the environment. This staging, however, is not an event, but an act of being "deliteralized." It is a demand to align oneself with the law and proclaim in self-ascription "Here I am!" Butler goes on to explain that the turn is an act conditioned by the conditions of the voice, as well as the responsiveness of the one responding to the law. Butler writes: "But where and when does the calling of the name solicit the turning around, the anticipatory move toward identity? How and why does the subject turn, anticipating the conferral of identity through the self-ascription of guilt? What kind of relation already binds these two such that the subject knows to turn, knows that something is to be gained from such a turn? How might we think of this 'turn' as prior to subject formation, a prior complicity with the law without which no subject emerges? The turn toward the law is thus a turn against oneself, a turning back on oneself that constitutes the movement of conscience." Judith Butler, *The Psychic Life of Power* (Stanford, CA: Stanford University Press, 1997), 107.

49 Fanon, *Black Skin, White Masks*, 109.

the resultant alienation is also seen to be elective by prioritizing external perceptions of the Black body, and any resultant psychic negation, over a more affirmative view of Blackness.

The above provocation can act as a foundation to think through Fanon's commitment to psychic actualization, and what this might mean in contemporary machine learning as an organizing practice of social fragmentation. More so, in what new ways could we consider how individual and collective consciousness is generated within inequitable relations, particularly as they collide with our visual organizations of the world?

Fanon challenges us to think beyond our initial perceptions of the racialized image by turning to theories of perception, namely those found in Gestalt psychology. Gestalt psychology, or configurationism, emphasizes how organisms perceive entire patterns or configurations as opposed to mere elements or individual components. The theory is useful in psychology to help determine how humans perceive visuals in relation to different elements or objects in the environment. An essential quality of Gestalt (translated from German as a "pattern," "figure," or "organized whole") is the notion that patterns can be configured and interpreted by humans as unified wholes.[50] The whole cannot be described as a mere sum of parts. What appears as a whole is in fact a perceived unity comprised of various distinct elements. These elements or objects may or may not be fully comprehended by the mind. But once the elements are registered, the mind works to refine them to construct a composite image. This process helps the mind make sense of an environment that is an otherwise disorderly and unstructured. Visual continuity also helps us replace objects that appear to be missing. We connect the missing pieces in order to reduce the complexity of the composite.

The notion of figure-ground organization is another component to Gestalt. Figure-ground organization is a type of perceptual grouping that is necessary for recognizing objects. It is

50 Robert J. Sternberg and Karin Sternberg, *Cognitive Psychology*, 6th ed. (Belmont, CA: Cengage Learning, 2012), 13.

also known as distinguishing a foreground or figure from a background. As such, in Gestalt perception is the relation between figure and background. The relation is causal, meaning that any variation in the figure will cause a topological variation in the perception of the ground. It is also reciprocal, in that any variation in the background will result in an altered perception of the figure. The brain uses perception to decide which attributes in the visual field are the figure and which are the background. In this sense, it is not unusual for the brain to alternate between perceptions of the figure and perceptions of the background.[51] In these terms, our brains assign meaning to certain attributes in the visual field that are linked to our perception of the inner world (*Innenwelt*) and outer world (*Umwelt*). The segregation of attributes helps us distinguish between attributes, as well as classify and organize them into coherent visual scenes. Put simply, in Gestalt what we see visually is not necessarily all that can be perceived. To perceive objects is, therefore, also the act of translating an object into perceptual attributes, such as shapes, sizes, and colors.[52] In the absence of unequivocal cues, it also brings forth any discontinuities in attributes or any disparities in depth, which can be interpreted as difference.

From Figure-Ground to the Color Line

Fanon directs our attention toward visual perception to help us understand how individuals with similar skin color can be perceived of as sharing a single racial narrative (or stereotype), despite an immediate recognition that each individual is a distinct being. Fanon contends that the projection of an anti-Black stereotype, and the subsequent assignment of meaning based on

51 In computer vision, the direction of the vision region is known as "border ownership," which determines any veridical correlation between object shape, size, or luminance and perception. See Tyler, *Computer Vision*.

52 Rudiger von der Heydt, "Contour-, Surface-, Object-Related Coding in the Visual Cortex," in *Computer Vision*.

this image, is a superficial argument. "When a 'blue' Negro—a coal-black one—comes to visit, one reacts at once: 'What bad luck is he bringing?'"[53] The collective consciousness in this example is not dependent on heredity but is the result of an imposition on the self-consciousness of Blacks that Fanon seeks to expose as an unconscious fantasy of white Europeans.[54] He warns that in the absence of reciprocity between racialized and non-racialized perception, Blacks are reduced to objects among other various inanimate objects in the world.

Fanon's proposal is motivated by Maurice Merleau-Ponty: in the "Eye and Mind" chapter of *Phenomenology of Perception,* Merleau-Ponty explores the art of viewing the world that is truly representative of what Leila Wilson describes as a "continuum of existence."[55] Wilson illustrates how, for Merleau-Ponty, the act of viewing with openness and immersion is in opposition with scientific thinking that fails to visualize through lived experience, or more specifically the body as the site through which one perceives, evaluates, and values the world. Perception, as Wilson notes, is not a channel that filters in and therefore reduces information from a separate environment. Instead, perception is an extension of consciousness that establishes a reciprocal connection between oneself and the world. In other words, the body's relation to the world is affectual, as it experiences not merely representations in its environment but its own movements within it. This type of kinesthetic awareness made possible by spatial equivalences is what Merleau-Ponty terms the "body schema."[56] The body is oriented spatially toward actual or potential actions. The body is in existence by "being toward" the world, where it projects its aims toward this goal.[57] Merleau-Ponty posits that the

53 Fanon, *Black Skin, White Masks*, 161.

54 Fanon, 162.

55 Leila Wilson, untitled essay on Maurice Merleau-Ponty's "Eye and Mind" (paper, Theories of Media, University of Chicago, Spring 2003).

56 Merleau-Ponty, *Phenomenology of Perception*, trans. Donald A. Landes (Oxon: Routledge, 2012), 128.

57 Merleau-Ponty, 173.

expressive unity between body and world is intentional, while equally active and receptive. Reciprocal expression is where communication is generated between the Self and the Other. It is also the site where the body and external world reorder each other in a "perpetual contribution" of reciprocal information exchange.[58]

Merleau-Ponty insists that all humans have a corporeal schema that determines how one positions themselves and acts in the world. The schema cannot be anticipated through a series of calculations. Human action is instinctive, and there is a "harmony between what we aim at and what is given, between intention and the performance."[59] The body is in a sense free to anchor itself in the world by establishing its own space within it. But it is Fanon's contention that this schema fails to account for Black lived experience. Fanon insists that the relation between the Black body and the world is not in harmony. Racism and the process of "negative socialization" is embodied in the corporeal schemas of the world. Racism, from this viewpoint, influences how whites are perceived, and how they perceive themselves as being in a space of superiority. Fanon's focus is on the act of embodied superiority, which is displayed negatively as a lack of recognition, or the will to recognize, the racialized body as being anything other than an object of deficiency.

Fanon contends that for Blacks, the corporeal schema does not promote authentic freedom. The schema is provided *for* him "by the other, the white man, who had woven [him] out of a thousand details, anecdotes, stories." Fanon asserts that racialized experience is more accurately described as an "historico-racial schema."[60] The historical draws upon the history of subjection (slavery, colonization, police brutality, and so on) to account for the persistence of racism and race-based violence in the world. While the racial or epidermal schema explains the process by which the perception of racial inferiority is internalized. The notion of the historico-racial schema is in direct confrontation

58 Merleau-Ponty, 100.
59 Merleau-Ponty, 144.
60 Fanon, 91.

with predominant ideas of human equitability within ontology. For this viewpoint, it is difficult to imagine a history of the world that is detached from the history of racism. Not to mention the role colonial imaginaries have played in the construction of the corporeal image. History, in this sense, is in a direct engagement with the fragmentation of Black lived experience, and not a continuum of experience as Merleau-Ponty claims.

Ultimately, Fanon's—like Merleau-Ponty's—concept of the schema reveals the relation between self-actualization, place, and temporality. While Fanon's schema suggests a universal language of control, historicity, and phenotype, it also argues against the phenomenological reduction of Black lived experience. His primary focus is more so on the dampening of psychic potential when the body is mediated through languages, perception, and visual recognition. It is not surprising, then, that he relies heavily on visual and artistic metaphors to illustrate the psychic effects of racial sorting and social categorization: "I had to meet the white man's eyes. An unfamiliar weight burdened me. The real world challenged by claims. In the white world the man of color encounters difficulties in the development of his bodily schema [...] I was battered down by tom-toms, cannibalism, intellectual deficiency, fetishism, racial defects."[61]

Fanon reminds us that perception is the site where the Black body is brought into the figure of the colonial imagination. In a direct critique of Gestalt, however, Fanon asserts that colonial perception is neither reciprocal nor reversible. Under colonialism, the object of Blackness can be simplified, fragmented, and brought into view. However, Black peoples cannot equally bring the non-Black other into any visible coherence. Pathologically, the resulting inequity blocks Black people from experiencing any direct view of the organizing principles of physiology, which is external and inaccessible to their perception. Keeling argues that although Fanon tries to move beyond the world of perception in *Black Skin, White Masks* toward a phenomenon of Black

61 Fanon, 83–84.

experience, he remains fixated on relocating Black being outside of the visual field. Keeling writes:

> This co-ordinate is always present in colonial reality, authorizing the continued existence of "the Black" and "the White" and posing a constellation of difficulties for the thinkers and makers of Black identity and culture. Those difficulties derive from the way that the spatiotemporal co-ordinate of colonization and enslavement is precisely what "the Black" indexes. But, it is also what must be wrenched upon and destroyed in order to liberate "the Black." Within Fanon's framework, that co-ordinate remains inaccessible. (For him, it is repressed.) Ontology thus becomes unthinkable. [...] The event of colonization and enslavement marks the violent creation of an antimony that is ceaselessly projected onto the world at the expense of "Being."[62]

Nonetheless, Fanon seeks to interrupt the temporal cycle of colonialism by disrupting the internal experience of the coordinate. To do so, he must shatter the formal legitimacy of the spatiotemporal coordinate and open perception to alternative forms of representation. Our non-veridical perceptions, in this sense, do not depend on any concrete sense of reality to make decisional judgments, even if the attributes perceived change over time. Following Merleau-Ponty, the percept, or the representation of what is perceived, is the simplest form of visual interpretation. Merleau-Ponty writes that the percept "is the simplest sense-given available to us," and is therefore the "very definition of the phenomenon of perception."[63]

Machine Learning and Gestalt

The link between sociogeny and machine learning is not immediately apparent. Psychic generation, particularly the pathological schema problematized by Fanon, offers further insight into the problem of race and Black psychic genesis, and how this might

62 Keeling, "'In the Interval,'" 98.

63 Merleau-Ponty, *Phenomenology of Perception*, 9.

be reinforced, even unwittingly, by machine learning applications. Sociogeny—as a context through which the generation of Black being can be conceptualized—might provide alternative ways to frame the contemporary relationship between race and machine learning.

In machine learning, computer vision techniques draw on Gestalt psychology to formalize figure-ground organization. The process helps facilitate perceptual groupings of data in artificial visioning systems. In computer vision, figure-ground organization refers to the detection of objects in an image, differentiating these objects from the background, and grouping contour elements into two-dimensional shapes that represent three-dimensional objects. It begins with an algorithmic interpretation of shapes, sizes, and colors of objects, and then applies the simplicity principle to recover the 3D shape. While it is recognized that 3D objects are conceptually complex, it is also known that objects never produce identically shaped retinal images. For instance, a table and a chair will never produce the same 2D retinal images, regardless of viewing direction. Each 2D retinal image would then be unique to a specific object; and therefore, any 3D shape can be recognized from its 2D representation. Computer vision is derived in part from the recognition that human observers cannot reliably reconstruct, or recognize, surfaces from depth cues. Although computer vision shares this impossibility, its primary task is to simulate the cortical architecture of human neurological function and to fill in details of 3D objects across regions that are missing in localized vision cues. (This is traditionally attempted by Bayesian techniques and other computational approaches.)

Fanon's exploration of the Black psyche can establish a consistency between sociogenetic process and the arresting of affirmative Black psychic genesis in machine learning. It is necessary to return to sociogeny, both its pathological and empirical relations, in any attempt at disrupting the durability of racial sorting and the negation of Black being in techno-racial ecologies. By techno-racial, I mean the materialization (and most often the production) of discriminations, prejudices, and reductions of life chances that emerge at the intersections of race and technologies that aid in the perception of Black peoples and the environment from a scientific point of view. The techno-social is key to the conditions that exist prior to any present engagements with data and machine learning models. The limits of Fanon's racial schema are inherent in the limits of linear causality, namely between racial categorization, machine learning, and the genesis of the Black psyche. This would require dismantling the schema in order to rebuild a new relation between the Black psyche and technology, which might constitute possible avenues toward the psychic liberation of racial alienation. This speculative potential between two interrelated objects (human and data) is consequential in the emergence of machine learning as it converges with the consistencies of racial logics.

ACT II
A Study on the Calculation of the Ordinary Racialized Object

CHAPTER 3
Machine Learning and a Thousand Manifestations of the Ordinary Object

> For you, it would be nothing but an indifferent picture, one of the thousand manifestations of the "ordinary."
>
> —Roland Barthes

In our current social and political climate, race and racism should be at the fore of any social commentary or discursive intervention. This has become even more important in the aftermath of the brutal execution of George Floyd by Derek Chauvin, a Minneapolis police officer, not to mention the continued killings of Black and brown people across the globe. And the digital landscape is no exception. The collision between the racial being and technology has long preceded the digital. The distances between our social relations are shortened through the prosperity of data-driven technologies such as machine learning, artificial intelligence, and the racial politics that constitute a critical foundation for the use of technology and its implementation. They also bridge the gap between the historical circumstances of race-based injustices and contemporary digital practices.

Today, the racialized being resides in a space of abstraction, between the digital and notions of technological progress. Technology flows through this space while rehearsing the languages of race and racism. Technology is a milieu that has been assigned the role of mediator between racial duress and discrimination. We must consider how race makes use of machine learning for the purposes of its own survival, particularly as it has yet to offer any substantial result that can account for the complexities of a racialized experience. While the algorithm, with additional training, might be able to recognize acts of racism, machine learning algorithms and their technocrats continue to misjudge the value of racialized life. Despite certain efforts—if any effort is made at all—the process of machine learning research and application is contaminated by a thick cloud of determinacy, which confuses the popular replication of racism for its value in the domains of language, the political, economics, sociality, and the law. The act of racism, as interpreted by the algorithm, is nothing more than a primitive engagement with the base levels of human violence. Machine learning's deficiency when it comes to understanding race as the decimation of human conviviality is at odds with the technocrat's valorization of complex mathematics as an objective and ahistorical phenomenon. Surprisingly, this point seems lost in the arrogances that have replaced genuine expertise, as if each code and keystroke is not bound to an already conditional set of values.

The exponential increase in the use of machine learning since the 1950s is equally astonishing, given its proliferation in nearly every public and private domain. But this is the precise objective: to gain insight into dynamic human behavior in order to reduce the levels of uncertainty that define all human and nonhuman life. It's an egregious claim that values humanity only inasmuch as the act of expressing human life can be put into service as a functional utility.

The operational objectives of machine learning are particularly important, but machine learning is not merely a functional apparatus. The mechanization of human life, as expressed by machine learning, exemplifies an obsession with the racialized principles of social order. It is here, at the intersection of race and governance, that machine learning gains its omnipotence. This is to say that machine learning gains relevance as the ceremonial placeholder for the white imaginary, whereby it gains power through the epistemic reinforcement of racial grouping.

In the Beginning There Were Obscure Definitions

What we know as machine learning today emerges from a number of distinct mathematical and computational methods, including applied and theoretical statistics, artificial intelligence, pattern recognition, and computer science. Nonetheless, definitions of machine learning are varied and broad in scope. Machine learning is generally defined as the study of data-driven methods designed to simulate, understand, and aid in human and biological information processing tasks. Tom Mitchell describes machine learning as a class of programs that improve through experience, and in doing so encompass types of problems that require sophisticated solutions.[1] David Barber, on the other hand, notes that the function of machine learning is not to mimic in exact replica of complex environments, per se. Instead, for him, machine learning seeks to enhance human decision-making through the rapid retrieval and processing of information into hypothetical and sometimes practical outcomes.[2] At the level of operation, machine learning can also be described as "the acquisition of structural descriptions from examples"—these examples being data "training examples" or "tuples," which facilitate the algorithm's ability to reach expected or unintended outcomes when given a sufficient amount of new or previously acquired data.[3]

These definitions are among the many that treat machine learning as a scientific or statistical practice rather than as an assemblage of human, technical, social, economic, and political processes. Recent engagements with machine learning, however, question whether machine learning can be understood in isolation. For example, Adrian Mackenzie advocates for a revision of the term "machine learning" in favor of the more inclusive term "machine learner," which includes the role of coder, engineer, civil infrastructures, organizational owners, and the public.[4] Mackenzie admits that his aim is not to insert more critical thought into machine learning research, but to think through machine learning as an environmental condition that incorporates the investment and practices of multiple stakeholders. Mackenzie argues that a focus on logical positivism within machine learning ecologies is a barrier to developing a more complete understanding of machine learning. Mackenzie draws on Foucault's critique of "positivity" to alert us to technical approaches that use empirical methods to justify exclusionary practices, while simultaneously obscuring the inner workings

1 Tom M. Mitchell, *Machine Learning* (New York: McGraw Hill, 2017).
2 David Barber, *Bayesian Reasoning and Machine Learning* (Cambridge: Cambridge University Press, 2012).
3 Witten, Frank, and Hall, *Data Mining*, xxiii.
4 Adrian Mackenzie, *Machine Learners: Archaeology of a Data Practice* (Cambridge, MA: MIT Press, 2017).

of their operations. Mackenzie asserts that "viewed from the perspective of control, and how control is practiced, machine learners perpetuate and epitomize the 'control revolution' [...] that arguably has, since the late 19th century, reconfigured production, distribution, consumption, and bureaucracy by tabulating, calculating, and increasingly communicating events and operations."[5]

The historicity of the control revolution is significant, in that our current appetite for a technocratic society highlights many recognizable features of historical exclusion. These exclusionary measures are not synchronic with our present anxieties. They occur diachronically as technologies change over time. In other words, machine learning is not a real-time image of the social, but a recurrent echo of the languages of Western discovery. As Foucault argues, these knowledges circulate through the unmitigated disclosure of hidden relations, coded into the agents of propositions that confirm or deny certain realities as being either true or false: "If a proposition, a sentence, a group of signs can be called 'statement,' it is not therefore because, one day, someone happened to speak them or put them into some concrete form of writing; it is because the position of the subject can be assigned."[6]

The statement of disclosure in machine learning, as expressed by claims of objectivity, is underwritten by a complex set of mathematical functions. These functions, although mathematically symbolic, are treated by the technophile as if they are impenetrable—self-justifying the protective shells created around these mathematics in order to preserve this "positivity of knowledge."[7] If these knowledges are articulated as an unmitigated discovery of hidden relations, coded into machine learning, then they invert the relationship between the assignment of proposition and the positionality of the computational

5 Mackenzie, 7.
6 Michel Foucault, *The Archaeology of Knowledge*, trans. Alan Sheridan (New York: Routledge, 2002), 107.
7 Mackenzie, *Machine Learners*, 6.

object. An assignment of algorithmic proposition puts us closer to Foucault's claim.

Foucault brings attention to the object of discovery as that which is considered desirable based on its practical use value. The mathematics that assist machine learning with the conversion of data into use value are underscrutinized, as are the algorithms that process the mathematical equations. Less considered are the hidden propositions that exist alongside the serviceable act of calculation in computational environments. One hopes that the technocrat might set themselves free from isolation and gain a greater awareness of the non-mathematical statements that are embedded in the very process of constructing an algorithm. This might help resist the inversion of the relations between engineer/algorithm and algorithm/end user, which are obscured by affirmations of knowing that are, in some cases, characterized by dogmatic assertions that conflate the replication of phenomena for proof of its existence or nonexistence.

Ultimately, it is the existence of human difference, and the process by which one claims authority over this difference, that must be considered in machine learning research—particularly in how the logical expression of racialized languages flow through algorithmic structures. Given the limits of Black recognition in the algorithmic, we must ask how, on the level of data and mathematics, algorithms might reinforce and in other instances reinstate the racial imaginary. To do so requires a deep understanding of the chondritic structure of machine learning, where the dynamism of life is predetermined by positivist expression.

This is an important entry point to approach a shaping of the world. Alternative ways of living can emerge if we commit ourselves to reframing the question of machine learning as one that is always already informed by racial classification and preemptive sorting. This does not require the disillusionment of retrospective occurrences of violence as the output of a learning algorithm, or of an externally driven ethical framework to establish workable boundaries. A commitment to the eradication of racism and other violence supersedes any efforts here, which begin with an integrated awareness of the symptoms of social ordering and the reinstatement of epistemic power.

To think through machine learning is to add multivalency to the perception of the Black technical object, as well as the process of machine creation. This would enact a return to the granular processes of computation that are centered on the functional operation of machine learning, as well as what a machine learning "problem" might mean given the various definitions of the field of research and its operational aims. In between these more macro- and micro-processes of machine learning is an oft-misrepresented relation between the data object (as conceived computationally) and the Black technical object. An elaboration on this relation might bring us closer to a type of radical transformation of the techno-human relation by providing new openings for alternative articulations of racial perception mediated by machine learning algorithms. The necessary shift is one bound by the ontological, and it promotes an alternative algorithmic praxis. To unearth this relation is to also recognize a pre-individuated capacity for praxis that might disrupt, dismantle, and rebuild the primal components of both racial and machine perception.

A "Non-mathematical Approach" to Learning Machine Learning

Michael Wood describes a "non-mathematical approach" to understanding the relation between the racial object and the historicity of machine learning, as well as the functional operation of the algorithms from which machine learning algorithms are derived.[8] This approach shifts focus away from a machine learning discourse based on a purely operational perspective, one that seeks an understanding of the relation between humans and computation as an act of code and mathematical equation. Instead, we should bring the complex realities of machine

8 Michael Wood, *Making Sense of Statistics: A Non-mathematical Approach* (Basingstoke: Palgrave Macmillan, 2003).

learning—including its intricate relationship with statistics, pattern recognition, artificial intelligence, and computer science—back into conversation with the logical theories from which the tools of machine learning derive, and more so the human-to-human relationships on which these theorems have been historically rehearsed.

While an understanding of mathematics is to some extent necessary if we are to undertake a more nuanced debate on the social impact of machine learning algorithms, to embark on a purely techno or social solutionism without due consideration of the parasitic relation between the two would be misleading at best. At worst, it leaves those invested in the production of a revised techno-human relation unable to consider the transference of racial anxiety as a certain object among other objects in the technical—and racial—imaginary. Notwithstanding, a journey toward a deeper excavation of algorithmic praxis, as well as the all-important historicity of machine learning and enumerative analysis on human populations, demands a reckoning with machine learning's inability to account for, let alone mitigate, racial perception—and, subsequently, rearticulate the act of solving the problem of the color line through numbers and quantities. By way of an alternative, machine learning in its earliest state, a mode of perceiving the world as a continuation of statistical activity, should be our starting point.

From Statistics to Machine Learning Algorithms

Despite common misconceptions, machine learning is not a new technology. What we know as machine learning today emerged from a number of distinct techniques, including artificial intelligence, pattern recognition, computer science, and statistics. Although machine learning and statistics are treated differently in intent and nature of analysis, it is difficult to distinguish between them since they share many core mathematical methods. Research in statistics and machine learning emerged in parallel, making it difficult to distinguish where machine learning begins and the field of statistics ends. In their 1984 text

Classification and Regression Trees, statisticians Leo Breiman, Jerome Friedman, Charles J. Stone, and R. A. Olshen introduce a technical schematic for the generation of decision trees for example problems.[9] In the 1970s and 1980s, computer science researcher J. Ross Quinlan was also actively designing a system to infer classification trees from data examples. Although the researchers produced their work in near isolation, and only became aware of each other's work years into their research, many of their methods for data classification (decision trees, regression, K-nearest-neighbor) were proposed as possible solutions to example problems in both disciplines.[10] Today, the same advanced classification methods are used extensively in statistics and machine learning research. For instance, Stone's and Olshen's regression is the most commonly used method in statistics as well as many machine learning algorithms.

While research in statistics and machine learning share some common methods, others are used interchangeably within both fields of research. For instance, in both statistics and machine learning, relationships between data populations and

---

9 Leo Breiman et al., *Classification and Regression Trees* (Boca Raton, FL: Chapman & Hall, 1984).

10 Statistical methods are typically delineated into five major categories of problems: regression, classification, supervised learning, unsupervised learning, and reinforcement learning. Regression is used to simplify complex data to express any mathematical relationships between two variables. From a learning perspective, regression helps simplify assumptions about mathematical structures, as well as any underlying generative processes within the data (for instance, hidden relationships between two variables). In mathematics, a variable is an alphabetic character that represents a number, or the value of the variable, which can be either arbitrary, unknown, or not fully specified. A range of simple and complex problems can be solved by treating a given variable as concrete. In basic mathematics, the solution to a problem can be achieved by replacing the variable with the respective coefficient of the equation. In more complex computations, however, variables are symbols that represent mathematical objects that can each be either a single numeral, a vector, or a matrix, which carry their own set of complex computations reduced to a single alphabetic character.

samples are modeled using symbolic mathematics. Nonetheless, statistical methods do sometimes enlist machine learning methods to test learning outcomes. Likewise, standard statistical techniques are applied to machine learning examples as data visualizations, attribute selections, and as methods for discarding outliers and simplifying statistical deviations. For example, machine learning models utilize both univariate and multivariate statistical methods, which are processed through complex algorithmic structures. These structures rely on abstract assumptions that are materialized into rule sets and circular input data for the purposes of knowledge production and computational insights. Furthermore, machine learning algorithms do not make assumptions about probability distributions. Instead, they rely on more inductive approaches to research. Machine learning seeks to build and iterate or adjust parameters in real time. Still, Jiawei Han, Micheline Kamber, and Jian Pei argue that it is often unrealistic and inefficient for systems to generate all possible patterns, so user-provided constraints and assumptions are seen as sufficient for research.[11]

These methodologies attempt to classify important indicators based on a given set of input variables. A classification problem can then be used to establish a set of tree-structured rules, from which decisions can be derived.[12] When the desired output consists of one or more continuous variables, the problem is labelled as regression. The robustness of these so-called decision trees is based on their ability to predict the "class" of sets of measurables. Additionally, Breiman, Friedman, Stone, and Olshen note that classifiers are not created informally, but are instead based on historical data and subjective experience. In their treatment of regression trees, the authors use data on

11 Jiawei Han, Micheline Kamber, and Jian Pei, *Data Mining: Concepts and Techniques* (Burlington, MA: Morgan Kaufmann, 2011).

12 Classification problems are cases in which the aim is to assign each data input to one or more of a finite set of predetermined categories. Depending on the problem, classification can be either those that produce accurate classifiers or those that uncover the predictive structure of the problem.

biological phenomena as learning samples to infer patterns within previously derived data. They use the example of a doctor who, through experience, is certain that an elderly patient with a heart condition can be safely classified as low risk if the person also has low blood pressure. Likewise, the residents of Los Angeles can be sure that a single hot day is likely to be followed by another. Although the researchers base their hypothesis on actual biological and social phenomena, these patterns are mediated through data and mathematical formulae. For instance, they describe the phenomena as extraction of certainty within an otherwise chaotic system. In reference to patterning, they state that a learned sample "consists of data on N cases observed in the past together with their actual classification," or more formally:

> A learning sample consists of data: $(x, j), \dots, (x_N, j_N)$ on N cases, where $x_N \in X$ and $j_N \in \{1, \dots, J\}$ where $n = 1, \dots, N$. The learning sample is denoted by $L = \{(x_1, j_1)\}, \dots, \{(x_N, j_N)\}$.[13]

The symbolism—even visually—is extraordinary in its reliance on historical data as a continuum of learned patterns in the future, which themselves are universalized into a generic relation between actual event and predetermined outcomes. According to Konrad Becker, classification is an assemblage of relations between general and universally objective attributions and subjective meaning. The assemblage also produces its own algorithmic reason that is obscured behind naturalized proofs. The tension challenges how we relate to data as the intermingling of the actual techno-social self and the scientific imaginary. Becker furthermore suggests that classification is not a sufficient strategy to discover accessible interrelations, yet it provides a conceptual space for knowledge production. According to Becker, these knowledges can reveal unseen problematics in the relations between humans and machines from within the sphere of unstructured data. Nonetheless, he warns

13 Breiman et al., *Classification and Regression Trees*, 5.

that knowledge mapping can be mistaken for what he terms "transient social fictions" that obscure facts, especially in terms of cultural difference. "It happens time and again," he notes, "particularly in relation to race, gender, social institutions, and any other domains where there is a vested interest in the making of realities."[14] Despite the symbiosis of statistical approaches and machine learning, Witten, Frank, and Hall distinguish machine learning as a process that digs through landscapes of available data in search of new hypotheses, whereas statistics in most instances sets out to prove hypotheses.[15] As mentioned, early machine learning algorithms aimed to learn representations of simple symbolic functions, which were then verified by experts. The intent was that the methodology would output a hypothesis that executed the correct or consistent classification of learning data. Consistent hypotheses, particularly verifiable ones, were problematic in that they relied on representations that may not have been computationally simple or easily verifiable. Additionally, training data often contained noise. This deviation from useable training data is particularly concerning, as it represents statistical efforts to normalize data deemed to diverge from the target function or solution.

Machine Learning and AI

As with statistics, machine learning has a lengthy relationship with artificial intelligence (AI) research. In fact, machine learning was not only considered a subfield of AI prior to emerging as a solidified body of research around the mid-1980s (although it is largely considered to have also derived from computer science and cognitive science disciplines), but was also a key technique in the overall development of AI. What distinguishes machine

14 Konrad Becker, "The Power of Classification: Culture, Context, Command, Control, Communications, Computing," in *Deep Search: The Politics of Search Beyond Google*, ed. Konrad Becker and Felix Stalder (Innsbruck: Studien Verlag, 2010), 164.

15 Witten, Frank, and Hall, *Data Mining*.

learning from AI, however, is a split in research aims. Unlike machine learning, as Pat Langley explains, "artificial intelligence and cognitive science were showing little interest at the time [mid-1980s] in learning-related issues."[16] AI and cognitive science emphasized more logic-based forms of knowledge production to derive measures of simulated intelligence. Early machine learning, on the other hand, was characterized by an emphasis on symbolic representations, such as mathematical logics and rules and decision trees. Still, although machine learning has moved away from its original research focus, it remained a subdomain of artificial intelligence until 2000, by which time it had fully separated from its parent discipline.

Today, although machine learning is considered a distinct body of research, a clear definition of the body of research is elusive. This is because machine learning approaches vary according to research and industry aims. That being said, Jaime G. Carbonell, Ryszard S. Michalski, and Tom M. Mitchell state that many machine learning techniques reside within three broad principles: (1) the underlying learning strategies used; (2) the representation of knowledge acquired by the system; and (3) the application domain of the system.[17] Langley, on the other hand, defines machine learning by its historical trajectory. He notes three major paradigm shifts in machine learning research that have altered the discipline and its relationship to knowledge production. According to Langley, early learning tasks prioritized transparency and clarity in learning methods, which were thought to open methods to experimental replication. He also outlines two more major shifts in subsequent research methodologies: First, early machine learning adopted an informal approach to small data sets. In early paradigms, such as explanation-based learning and theory revision, researchers

16 Pat Langley, "The Changing Science of Machine Learning," *Machine Learning* 82, no. 3 (March 1, 2011): 275.

17 Jaime G. Carbonell, Ryszard S. Michalski, and Tom M. Mitchell, "An Overview of Machine Learning," in *Machine Learning: An Artificial Intelligence Approach*, ed. Ryszard S. Michalski, Jaime G. Carbonell, and Tom M. Mitchell (Berlin: Springer, 1983), 3–23.

preferred using fewer training data sets, while limiting the use of background knowledge to drive the learning process. Researchers attempted to solve more complex problem sets (in more unstructured systems) by distancing the learning process from a reliance on input data and human expert knowledge. In doing so, artificial self-learning could mimic the cognitive capabilities of humans more closely. Langley argues that a shift toward self-learning required a more formal approach that emphasized machine optimization and performance in specific complex systems. This methodology is still in use today, where it finds practical value in a variety of scientific domains, particularly in poorly understood domains where humans previously lacked the knowledge needed to design more complex algorithms (such as facial recognition), as well as domains that required different scales of adaptability to dynamic contingency—in self-driving automobiles (although, they are also confined by rule-based logics), search engines, and automated manufacturing settings, for instance.

Although machine learning algorithms depend on data, previous advancements in research were limited by the availability of data sets. However, early research accelerated with increases in the volume of accessible and open data sets. By 1987, with the availability of more open-access data, research moved beyond pure experimentation. Researchers were no longer limited to experimenting with small self-aggregated data sets with predefined learning problems, but could mine more extensive unstructured data. They could derive their own learning examples to test the capabilities of more dynamic learning algorithms. Prolific data mining activities increased the ambitions of research activities. For instance, in 1987 University of California, Irvine, PhD student David Aha aggregated vast amounts of data into what became known as the UCI Machine Learning Repository. The repository was not only open source, but was also simplified, easier to measure against learning examples, and accessible via FTP. Langley argues that although data accessibility contributed to advancements in machine learning research, it also resulted in the oversimplification of learning research, enacting an over-reliance on component learning algorithms, as well as an increase in research conducted in isolation from wider, more large-scale intelligence systems.

Consequently, machine learning was seen less as an exploratory science than as a strictly performance-based method for data classification. By the mid-1990s machine learning had completely withdrawn into more "complex" tasks like multistep reasoning, heuristic problem solving, and language understanding that had defined earlier movements. This new direction was thought to foreclose on initial exploratory methods. For prominent researchers, like Langley and Herbert A. Simon, performance is measured by classificatory accuracy and the speed at which defined problems are solved.[18] Thereafter, machine learning research was seen as providing little value if it was not aligned with "effective" (read, measurable) learning performance. Under this new objective, analysis was carried out in a number of ways, which typically involved worst-case scenarios and systems-error rate analyses and assessments of reaction times. In all, these methods suggested new ways of structuring learning problems.

Machine learning research, as such, was centered around three primary methods: (1) task-oriented studies, or the engineering approach (the development and analysis of learning systems to improve performance in a predetermined set of tasks); (2) cognitive simulation, with an investigation of computer simulation of human learning processes; and (3) theoretical analysis, or the theoretical exploration of the space of possible learning methods.[19] Under this rubric, early researchers sought to scientifically prove the validity of computational learning problems. The primary aim was to assess whether or not a specific learning problem was desirable for further investigation. Researchers adopted informal, exploratory analytical approaches by using innovative methods to demonstrate possibilities of learning

---

18 Pat Langley and Herbert A. Simon, "Applications of Machine Learning and Rule Induction," *Communications of the ACM* 38, no. 11 (November 1995).

19 Carbonell, Michalski, and Mitchell, "An Overview of Machine Learning," 1–4.

outcomes. More specifically, the overall determinant of a successful outcome was measured by the algorithm's ability to learn through data and human-aided exploration.

Whereas earlier analytic logics were underwritten by knowledge-based expert systems, the above factors were later thought to be best achieved by measuring the algorithm's performance with practical and real-world problems. In his signature work on learning and decision trees, Quinlan details the discipline's initial reliance on known or explicit data sets to drive system outcomes. These systems typically required an immense number of rule sets to establish the learning environment and develop a context by which performance could be measured. Quinlan notes that despite the volume of rule sets necessary for training, computational limitations forced engineers and specialists to code environmental rule sets by hand.[20] Quinlan uses Mitchell Feigenbaum's description of early research as a "bottleneck" in the operation of expert systems, which Feigenbaum later argued could be more readily solved if the discipline shifted its focus to strategies of machine performance rather than exploratory capability.

Quinlan believed machine learning was far more than a craft or praxis. He thought the capacity of machine learning would be more fruitfully realized if it were trained to explicate knowledge as a method for computational learning. In the 1980s, research on machine learning diverged from symbolism and representation to include methods in pattern recognition, as well as probabilistic and instance-based representations. These methods included the incorporation of ideas borrowed from the field of pattern recognition, which had already been developing methods on statistics-based performance. According to Quinlan, application domains, also known as evaluation environments, find validity in applied areas (or statistical spaces that involve classification).[21] Domains are characterized by representations

20 J. Ross Quinlan, "Induction of Decision Trees," *Machine Learning* 1, no. 1 (March 1, 1986): 82.

21 Quinlan, 82.

of acquired knowledge through decision rules or "trees."[22] These systems are rudimentary and lack the expressive power of more complex systems, like semantic networks or other first-order representations. For this reason, they are considered less capable of learning in more dynamic and contingent ecologies. Furthermore, decision trees require inductive methods to generate hypotheses. Here, domains are seen as sufficient enough to produce knowledge by aligning problem solving with predefined practical applications. Less considered is the reduction of the conditions for decision-making, which in the falsification of hypotheses assume a certain passivity of dynamic phenomena. The domain is presupposed as an always already closed environment. But any derivation of knowledge by artificial decision is statistically moderated by technological efficiency. In other words, decision-tree systems function in isolation, giving the appearance of autonomy.

As such, an impasse developed between early analytic and heuristic ambitions and more performance-driven praxis that could self-infer classifications from training examples. Consequently, design technique shifted from praxis as enrichment to praxis as performance, which meant the operation of the algorithm was deemed successful only inasmuch as it could accurately regurgitate any given training data—much like the shift in aims in early statistical data aggregation. Dynamic mathematical formulae, championed by Leibniz and later augmented by the work of George Boole and others, helped researchers develop a universal language and formalized set of rules that could obviate the necessity of human participation in the computational process, increase efficiency, and relinquish the bottleneck problem. Leibniz invented differential calculus under the principles of logical additions, subtractions, multiplications, and divisions into

22 Decision trees are defined as "graphical representation of a decision table showing possible paths and actions if different conditions are met." John Daintith and Edmund Wright, eds., *A Dictionary of Computing* (Oxford: Oxford University Press, 2008), 83.

an arbitrary form of potentials—methods from which artificial intelligence would capture neural activities through simulation.

Trained Inferences, or How to Train a Racist Algorithm

In many instances, machine learning algorithms are assumed to derive knowledge or patterns from structured and unstructured data that is often considered "raw" and uncompromised.[23] If engineers are aware of the various fallibilities in input data, they rely on the mathematics that underpin machine learning to mitigate and "clean" any suspected data. Importantly, machine learning objectives are driven by computational processes that best fit data to established (or unspecified) hypotheses. These hypotheses might be specified by engineers or, in the case of advanced machine learning models like artificial neural networks, might function as part of their own computational understanding of environmental phenomena. These inferences are sometimes heuristic or based on observable data patterns and prior knowledge held by engineers. In other instances, they are inferred from computational perceptions of the environment.

Machine learning algorithms are comprised of a sequence of algorithms, which are themselves an array of step-by-step mathematical instructions. Most approaches to machine learning are, however, organized around a two-step process: learning (or training) and classification from the data. In the learning step, algorithms are coded to accumulate and learn from training data. Training data typically consists of "tuples" (also known as samples, examples, instances, data points, or objects). Tuples constitute database records and are initially derived from previously aggregated data sets. Tuples are also assigned to output classes. Once a tuple has been given an identity (or a specific class), it is referenced by the algorithm as input data. The algorithm is

23 See Lisa Gitelman, ed., *"Raw Data" Is an Oxymoron* (Cambridge, MA: MIT Press, 2013).

then instructed to classify the tuple based on its initial input. For instance, an input tuple:

> var named = (E, "Enrolled");
> var named = (P, "Fees Paid");
> should match the output classification:
> var unnamed = (Answer: E + P, Message: "Student")

To achieve successful classification, researchers use various types of training methods, depending on performance objectives. In supervised learning, tuples are "supervised" or forced into appropriate output classifications. With unsupervised learning (or clustering), output classes are left unlabeled. In unsupervised learning, algorithms are instructed to make best guesses and repeat the process until tuple classes are understood. Unlike supervised examples, unsupervised examples are more universal, since prior classifications have not been determined. This is beneficial in problems that require the grouping (or clustering) of similar data. This is particularly true in situations where data in complex environments needs to be mapped onto two- or three-dimensional visualizations, or to gain deeper insights from database records. Another widely used learning method is reinforcement learning, which is a beneficial technique in instances where decision-based actions are necessary. Reinforcement learning algorithms discover the most optimal outcome by trial and error, in contrast to a predefined set of outcomes in supervised learning. However, complications arise when output quality can only be assessed indirectly or longitudinally before any clarity of results can emerge. If the classifier has a high success rate, then it is used on future unknown (or previously unseen) data. For example, data from previous student applications—race, socioeconomic background, place of residence, and so forth—can be used to predict rates of retention for future students.

In complex environments, target functions are an important formula in machine learning algorithms. The target function is the formula that an algorithm feeds data to in order to calculate predictions. It is useful for real-world machine learning applications where—unlike training sets—outputs are unknown, or where accurate performance is difficult to replicate outside of controlled environments. In controlled environments, performance accuracy is likely to be higher than in real-time results. By analyzing large quantities of data related to a specified problem, an algorithm can detect consistencies in the data, while also gaining understanding of previously unspecified rules about how the subject under study operates. The success of this operation determines the learning capability of the algorithm, which is fed multiple tuples in multiple cycles until desired rates of successful classification are achieved.

According to Han and Kamber, once an algorithm is trained in a complex environment, new rules and patterns can be established to either classify or preempt similar phenomena.[24] For example, financial institutions might use a data model to solve the problem of "creditworthiness" by classifying consumers based on income, previous loan acceptance, immigration status, length of residence, or other variables that are understood to influence the likelihood of default. Based on analysis, a number of outputs are possible: "accepted with x or y% interest rate," "denied," or "more information needed." Similarly, a national security agency that has established the problem of "radicalism" might use facial recognition, social media posts, and other behavioral data to determine what persons might be deemed "suspicious," "a threat," or "sympathetic to radicalism against state power."

Successful Data

Generalization is defined as the successful categorization of data training sets. However, generalization is challenging in practice, as data sets are far too vast and variable. Furthermore, training inputs might only represent a small fraction of all possible data available. The ability to generalize is central to recognizing patterns within the data, and is seen as necessary to make sense

24 See Jiawei and Kamber, *Data Mining*.

of aggregated information. One such application is face detection in high-resolution video streams. Here, the computer must handle large amounts of data in numbers of pixels per second, input them into complex pattern-recognition algorithms, and subsequently categorize the output. The feasibility of this operation depends on a number of variable factors, such as hardware limitations and system parameters that affect computational speeds. Christopher M. Bishop cautions against the discarding of pre-processing information that is important to problem solutions.[25] This removal, either accidental or intentional, can reduce the overall accuracy of the system. However, research has shown that by making use of unsupervised methods as opposed to supervised data sets, machine learning can produce accurate facial recognition with high levels of sensitivity to non-targeted objects.

Nello Cristianini and John Shawe-Taylor argue that as machine learning practitioners become increasingly interested in the particular roles of training data, quality becomes more difficult to measure in the field, leading to a deeper problematic. For instance, even if a hypothesis can be found, it may not make correct classifications of unseen data. For this reason, attempts at optimizing the learning algorithm are made to more closely class, or generalize, data not found in the data set. Generalization is in a sense, then, a search for a "best" match according to a set of examples. Cristianini and Shawe-Taylor note that shifting the focus of machine learning to generalization removes the need to view hypotheses as correct representations and instead as problems of optimization. The shift reduces the potential of hypothesis exploration by simplifying the satisfaction criteria for successful learning: if the hypothesis produces the right output, it satisfies the generalization criteria, which then becomes a functional measure by which certain data is classified—as opposed to only description of it as a classed set. Cristianini and Shawe-Taylor write that "in this sense the criterion places no

25 See Christopher M. Bishop, *Pattern Recognition and Machine Learning* (New York: Springer, 2006).

constraints on the size or on the 'meaning' of the hypothesis—for the time being these can be considered to be arbitrary."[26]

Despite the above, generalization should not be disregarded entirely, as it aids in the reduction of large rule sets. Generalization helps reduce error and redundancy, thereby increasing overall efficiency and decreasing processing times and operating costs. The fact that an algorithm is based on statistical limits can be seen as a motivating factor for algorithms to optimize within certain parameters. However, the approach loses robustness in that an algorithm is only as good as the results that motivate its optimization. Researchers argue that the strength of this paradigm is in its reliance on proven statistical science and not on previous heuristic approaches that may be based on "misleading" intuition. Of primary importance are attempts at scaling complex data sets, which range from small academic ones to large, so-called realistic data comprised of hundreds of thousands of examples.

Kenneth Cukier and Viktor Mayer-Schoenberger suggest that aggregating large amounts of data is preferable over smaller samples, and that by accepting the "messiness" of data and "giving up" on known causes, sampling error is not only reduced, but insignificant.[27] They argue that a sampling error in big data sets is a signpost for a truth, since the immensity of the data set is such that it encompasses all possible occurrences. To the contrary, Cathy O'Neil and Rachel Schutt remind us that samples never contain everything that is available, regardless of size, and are "often missing the very things we should care about most."[28] O'Neil and Schutt use a counterexample of election polls. They illustrate that even if every single person who

26 Nello Cristianini and John Shawe-Taylor, *An Introduction to Support Vector Machines: And Other Kernel-Based Learning Methods* (Cambridge: Cambridge University Press, 2000), 4.

27 Kenneth Cukier and Viktor Mayer-Schoenberger, "The Rise of Big Data: How It's Changing the Way We Think About the World," *Foreign Affairs* 92, no. 3 (May/June 2013): 28–40.

28 Cathy O'Neil and Rachel Schutt, *Doing Data Science: Straight Talk from the Frontline* (Sebastapol, CA: O'Reilly, 2013), 25.

leaves a polling station were sampled, it would fail to capture the decisions of the people who decided not to vote that day. They warn that paradigms of omnipresent data collection exclude the voices of people who don't have the time, energy, or access to cast their votes. They warn that these people then become invisible to the process, skewing the results. The importance of these exclusions may not be apparent when considering, say, Netflix recommendations. They could, however, produce more significant consequences in other contexts, such as the impact early predictions may have on late voters.

Under the assumption of ubiquity, data is often translated as being objective or "speaking for itself." O'Neil and Schutt note that working from within this assumption can lead to false beliefs about actual social practices and may lessen the impact of other discriminatory forces that inform observed patterns. In other words, they write explicitly that "ignoring causation can be a flaw, rather than a feature. Models that ignore causation can add to historical problems instead of addressing them. And data doesn't speak for itself. Data is just a quantitative, pale echo of the events of our society."[29]

How Does It Feel to Be a Target Problem?

Machine learning algorithms approximate target functions from noisy data, which is commonly thought to negatively impact performance. This is significant, considering the general understanding of noise as data deemed irrelevant or inconsequential to a computational problem at hand. While some algorithms are sophisticated enough to detect noise, they are sensitive to underfitting and overfitting. Underfitting occurs when the learning algorithm can neither model the training data nor generalize new samples for data representation and classification. This is particularly useful in environments that are overly complex, where data aggregation is too expensive, or in cases where the

29 O'Neil and Schutt, 26.

best methods to approach a problem are unknown. Underfit models tend to have poor performance, for example, when applying a linear model to nonlinear data. In this case, the model is too simple to explain the variance in the data, resulting in higher degrees of bias. Overfitting, on the other hand, takes place when a target function is too closely fit to a limited quantity of data, resulting in a model that matches too closely to the error within it. Overfitting is problematic when attempting to find patterns or develop elaborate theorems from the data, as they might reduce predictive power with less accuracy. With overfitting, what might appear to be a pattern in the data might in reality be a chance occurrence, particularly when the model is applied to real-world problems.

Data analysts have long addressed the problem of underfitting and overfitting though temporal mathematics. Statistically, although data might be overfit or underfit, errors or distortions in present data sets can be used as intelligence for future models. In these cases, emphasis is placed on historical data sets. New thresholds are then placed on the hybrid past-present data to account for present error. Temporal manipulation allows for speculation on any number of observable or nonobservable gaps in the data. Variables, in these cases, are treated as if they are explicit numbers. By treating variables in this way, a range of problems can be solved in a single calculation. In basic mathematics, a simple problem can be solved by replacing a variable with its respective coefficient of the equation. In more complex calculations, however, variables are symbols that represent mathematical objects—either a single numeral, vector, or matrix—that carry their own set of complex computations reduced to a single alphabetic character.

Gaining control over variability is a crucial exercise for data analysts who seek to reduce negative risk factors. For this, analysts might turn to reinforcement learning or continuously valued learning algorithms. Reinforcement learning is useful for the assignment of threshold values, which can be thought of as a boundary or line that separates data of interest from "othered data," or data deemed inconsequential for a specific computational problem. Othered data are variable inputs, such as the fragments of human behavior that live outside the desired population. Depending on the established problem, a desired population might be measured in terms of an acceptable rate of retention, or in some cases, exclusion from the process. In the former, to achieve a more desirable population, institutions might aim to reduce data deficits, as mentioned above. This would mean that those in the population who are not aligned with thresholds of achievement or "satisfaction" might be singled out or targeted for future action.

Unlike many notions of exclusion (like clustering and classification, which seek to single out data based on either a categorical or discrete set of characteristics), reinforcement learning is an area of machine learning concerned with the assignment of cumulative rewards for actions taken in an environment. Reinforcement learning enables the generalization of continuous variables in the data, which are variables that can keep counting over time. Continuous variables can be infinite. Variables can also take many shapes, as long as the values can be assigned measure. A discrete variable could be the number of university students (8,523), whereas a continuous variable could be a person's age (17 years, 8 months, 3 days, 6 hours, 12 minutes, 2 seconds, 1 millisecond, 8 nanoseconds, 99 picoseconds, and so on). Accounting for continuous variability is advantageous in reinforcement learning, since variables require less specificity in more dynamic environments. From a modeling perspective, reinforcement learning helps simplify assumptions about data structures. They hold the statistical power to identify hidden relationships between two variables, which can be arbitrary or specified. As a result, reinforcement learning models are less dependent on the accuracy of variable quantities. Instead, they focus on finding a balance between exploiting current knowledge and exploring unknown patterns or correlations. Reinforcement learning is well suited for problems that include long-term versus short-term reward trade-offs or risk assessments (for instance, timetabling, space allocation, student recruitment targeting in the university environment, or robot control, telecommunications, and AlphaGo in computational environments).

Recognizing Patterns

As with some statistical methods, machine learning also adopts methods for probability modeling. Despite variability, models rely on the recognition of patterns in data. This assists in the production of representations, such as spatial and temporal data visualizations and multimedia or textual graphs. In statistics, pattern recognition models are crafted around dynamic rule sets, and are concerned with the discovery of regularities in data, which are then classified into discrete classifications. Tuple classifications are known in advance, and are sometimes specified after initial training and inspection. Pattern recognition may also utilize machine learning approaches to optimize the function of adaptive models; and in many cases, tuples are condensed and simplified to increase algorithmic efficiency and processing speeds. Pattern recognition models use machine learning techniques to increase the field of vision by subsuming randomness, or the contingency of variation, into a generalized group of patterns.

An essential function of pattern recognition is the ability to make "nontrivial" predictions from new data sets. Data can also be modulated between black-box patterns (structures that are incomprehensible and thus hidden from view) and transparent-box patterns (which reveal explicit structures). It is assumed that key insights can be inferred from either method. However, Han and Kamber note that not all inferences are of interest to researchers. They argue that this raises serious concerns for data mining, and that for a pattern to be "interesting," it must generate knowledge that satisfies the following criteria: (1) the pattern is easily understood by humans; (2) it is valid on new or training data with some certainty; (3) it is potentially useful; and (4) it is novel.[30] Han and Kamber's appeal to interest as insight reveals the fragility of machine perception as a function of human desire and expectation. They blur the relation between human and machine as an activity of self-interest diluted under

30 Han and Kamber, *Data Mining*, 21.

the logics of comprehension. Pattern recognition models, as with similar approaches, are what brings the Other into view.

A powerful component of this method is its sensitivity to the relations between seemingly independent objects and variables. Take Stochastic Pattern Theory (SPT), a revised approach to pattern recognition that identifies the hidden variables of a data set using real-world, real-time data rather than artificial information. SPT champions David Mumford and Agnès Desolneux draw on Ulf Grenander's groundbreaking work in pattern theory to argue for a move away from traditional methods in pattern recognition to account for what they call meta-characteristics.[31] This approach is applicable to facial recognition and other algorithms where recognizable patterns are desired. These meta-characteristics are motivated by the identification of cross-signal patterning that correlate distinct patterns or signals within either single or multiple systems. While data denotes the information contained within a signal, signals represent the way we communicate (for instance, by speech, a transaction, where we travel and how often, what we read, a hand gesture, or other body language). SPT models are thought to improve operational efficiency by offering, say, a security agency a comprehensive study on the random movements of their targets. For service organizations like higher-education institutions faced with "chaotic" student behavior, SPTs promote operational efficiency. Although distortions or misrepresentations in stochastic models are difficult to trace (for instance, when assigning a smile, grin, or grimace to an individual's face), Mumford and Desolneux argue that by modeling the cross-pollination of distinct attributes, one can derive practical insights from what may initially appear to be noise or variability. As such, an SPT is less reliant on error in the present, as any distortion in results can be used to improve future stochastic models. In this way, predictive models become champions of economies of risk and future scenario building.

31 David Mumford and Agnès Desolneux, *Pattern Theory: The Stochastic Analysis of Real-World Signals* (Natick, MA: A K Peters, 2010).

They are staged as the vanguards of population control reinforced by the power of mathematical assumption, where the generalization of discrete desires, behaviors, and movements are put into service of capital or operational gain.

Statistical Pattern and the Object-at-Risk

The visibility of the racial being as data is dependent on the modulation of particular attributes for the calculation and subsumption of these attributes into universally accepted patterns of behavior. The normalized pattern then takes on the role of a threshold value by which the terms of access are both regulated and enforced. This abstraction (and particularity) distances the body from actual corporeal being and turns it into a body as risk and capital value. Here, the body is visible only inasmuch as it subscribes (willingly or not) to the objectives of capital growth.[32] Whereas these terms were most readily negotiated on the level of the body—the bioepistemic transfer of labor and death and violence as risk itself—machine learning has rearticulated the means by which capital and the logics of colonialism move further into the means of discrete efficiency. That

32 Recent advances in machine learning have led to increased concerns over the impact machines will have on workers and work environments. As machines learn to perform more complex tasks, questions arise as to what roles machines will play in future decision-making, in both enterprise and everyday environments. Despite the recognized benefits of AI and automation on productivity and in reduced human toil, issues arise over how the accelerated growth of machine capabilities and automated work environments might result in increased job loss, or the so-called hollowing out of what we know as work today. As result, more medium- and low-skilled jobs are at risk of being made redundant by automated processes. It is estimated that within two decades 60 percent of work currently being performed will be entirely replaced by machines. Others argue that because machines lack the ability to tackle simple human tasks, we should shift our focus toward ways to gain the highest value from human-machine interaction.

said, it is no coincidence that machine learning is put to use as the mediator between capital, the body, and the discrete, as the operation of patterning is embedded within the foundation of machine learning research.

To enable data representations (spatial, temporal, multimedia, or textual), statistical pattern recognition approaches are applied for classification. Pattern recognition is concerned with the discovery of regularities in data, which are then classified into discrete categories. Douglas H. Fisher suggests that the first pattern recognition algorithm emerged in 1936, with the goal to build an autonomous machine that could process a variable set of input data and generalize them into a patterned output.[33] In pattern recognition models, tuple classifications are known in advance, which are sometimes specified after initial training and inspection. As in machine learning, pattern recognition models are crafted around dynamic rule sets. Pattern recognition may also utilize machine learning approaches to optimize the function of adaptive models; in many cases tuples are condensed and simplified to increase algorithmic efficiency and processing speeds. Pattern recognition researchers use machine learning techniques to increase the field of vision by subsuming movement, or the contingency of variation, into a comprehensible generalization. A powerful component of this method is its sensitivity to the relations between independent objects. These objects may or may not be grounded in anything other than abstraction. For instance, observed similarities in skin color between human populations and animals, associations of mundane behavior and security threat, and so on.

33 Douglas H. Fisher, "Knowledge Acquisition via Incremental Conceptual Clustering," *Machine Learning* 2, no. 2 (September 1, 1987): 139–72.

Confronting the Limits of Perception

While this brief act of extracting the specificities of machine learning in function does not account for the complete array of technical capacities, the obscure and often contradictory definitions of machine learning reveal the fragility of any claim to its general direction, purpose, or perceived benefit, which is always already in balance between the objectives provided to the algorithm, our release of experience into a machinic interpretation of perception, and any claim to technological objectivity. With this starting point we are faced with a crisis of artificiality. But this crisis is not one that insists on additional computational capability to know that which is already known in our present human-to-human relations; it is rather a crisis of the self in relation to the perpetuation of fictive aims for the illusionary betterment of a dynamic of race that is barely understood outside of those who are in direct experience, along with a technology that is itself barely comprehended beyond what is instructed and prescribed. While we might want to believe in the ceremony of machine learning as a key to the future, we ourselves are the ones structurally responsible for technologically mediated judgmentsof the world. If machine learning orients us toward a future whereby accountability for our relations is wholly shifted into that which can only pattern and reduce the dynamics of race, then our aspiration toward the beyond and unknowable becomes no more than a rudimentary understanding of the world that has predetermined the fragmentation of culture. To challenge this crisis, our task is to embark on a fundamental reevaluation of the cultural values that are always already informed by our individual and collective perceptions of the world and each other. We must bring to light the notion of a new aspiration that resides beyond the threshold of existing racial perception, and instead release our commitment to the fragmentation and reassembly of difference itself.

CHAPTER 4
Neural Mimicry and Early Learning through Race Play

Historicity and the technical function of machine learning necessitates a reformulation of techno-human relations. In many ways, machine learning operates within a human capacity to replicate existing racial violences imposed by the strains of racial perception. If we think of the spatiotemporal order of machine learning and race as a continual process of racial reinscription, then we find ourselves within a cycle of alienation, whereby a new more equitable planetary relation can be imagined only inasmuch as it remains bound by that which has already proven itself inadequate for the task. Diving deeper into the relational network between machine learning and other data and computational sciences helps us gain new views on what can be seen as the various intentions of machine learning advancement and the development of early algorithmic techniques for the purposes of racial discrimination.

Simulating the Mind: Computer Science and Neural Networks

Along with its parasitic relation to statistics and artificial intelligence, it is necessary to understand machine learning as a subfield of the wider discipline of computer science. As with machine learning, definitions of computer science (sometimes called "informatics") vary widely within research and industry. In the sciences, emphasis is placed on formal knowledge building. For instance, Norman E. Gibbs and Allen B. Tucker first defined the field of computer science as the study of algorithms, including their formal and mathematical properties, hardware realizations, linguistic structures, and applications.[1] In contrast, early computer scientists Allen Newell and Herbert A. Simon see computer science as an empirical discipline, or more specifically, a living machine that extends beyond the experimental sciences.[2]

1 Norman E. Gibbs and Allen B. Tucker, "A Model Curriculum for a Liberal Arts Degree in Computer Science," *Communications of the ACM* 29, no. 3 (March 1986): 204.

2 Allen Newell and Herbert A. Simon, "Computer Science as Empirical Inquiry: Symbols and Search," *Communications of the ACM*

For Newell and Simon, as with Erol Gelenbe and Jean-Pierre Kahane, the study of algorithms is both a technology and science.[3] Newell and Simon, however, see computer science as an essential tool to deepen our understanding of nature. In this way, they argue that the field can be viewed as the study of organisms rather than the study of code and mathematics. Gelenbe and Kahane, on the other hand, posit that computer science is closely aligned with natural sciences, such as physics and engineering. The purpose of the field, according to them, is to build models that can form relationships between observable phenomena and scientific measurement, and ultimately to formalize the "embodiment of concepts into [scientific] artifacts."[4]

The definition of computer science is equally as broad in industry applications. For example, human-centered technology and information security specialist Angela Sasse defines computer science in terms of its capital value, while Derek Hill and Sylke Grootoonk define the field by its computational efficiency, as well as its capacity to solve difficult scientific and industry problems.[5] Problems can range from system algorithmic efficiency and hardware design to code and language execution. Computer vision, facial recognition, as well as speech and language processing, are key applications of computer science techniques.

Most computer science models in both research and industry adopt deep learning approaches. Deep learning, developed by Geoffrey Hinton and colleagues from the University of Toronto, along with Li Deng and Dong Yu, is a subfield of machine learning concerned with neural-network algorithms inspired by the

---

19, no. 3 (March 1976): 113.

3 Newell and Simon, 19.

4 Erol Gelenbe and Jean-Pierre Kahane, preface to *Fundamental Concepts in Computer Science*, ed. Erol Gelenbe and Jean-Pierre Kahane (London: Imperial College Press, 2009), v.

5 Derek Hill and Sylke Grootoonk, "Disease Management: Biomarkers to Digital Health," *European Pharmaceutical Manufacturer Magazine*, October 22, 2015, https://www.epmmagazine.com/pharmaceutical-industry-insights/disease-management-biomarkers-to-digital-health.

structure and operation of the brain. Andrew Ng describes deep learning as "our best shot at progress toward real AI"; as such, the idea of deep learning is to use brain simulations to make algorithms better and easier to use.[6] Whereas early neural networks lagged behind SVMs in performance, according to Ng, larger, more efficient neural networks can now be trained thanks to faster computers and access to larger data sets.

Deep learning is a hierarchal feature learner, meaning the technology can perform automated extraction from raw data. Using feature learning, deep learning exploits unknown structures in input data in order to discover representations at multiple levels. Deep learners can thus produce more nuanced and elaborate representations at higher levels of abstraction than had been previously achieved. Importantly, automated-feature learning allows the system to learn complex function mapping (of input data to target output) directly from the unstructured environment, without relying completely on human intervention.

Google Brain is one of the earliest and most prominent examples of deep learning. Google Brain is networked through over one thousand computers coded with over one million simulated neurons and one billion simulated neural connections. Google Brain is used extensively throughout Google's portfolio of products, and is used to accumulate and classify large amounts of unsupervised user data, which can be shared between products and services. Strikingly, three days after its first iteration Google Brain had mapped ten million still images from YouTube, revealing recurring patterns of human faces, human bodies, and cats. Quoc V. Le and his colleagues note the model's successful advancement of facial recognition and computer-vision research: "Contrary to what appears to be a widely-held intuition, our experimental results reveal that it is possible to train

6 Andrew Ng, "Deep Learning, Self-Taught Learning and Unsupervised Feature Learning" (lecture, Institute for Pure & Applied Mathematics, Graduate Summer School 2012: Deep Learning, Feature Learning, University of California, Los Angeles, July 10, 2012), https://youtu.be/pfFyZY1RPZU.

a face detector without having to label images as containing a face or not. Control experiments show that this feature detector is robust not only to translation but also to scaling and out-of-plane rotation. We also find that the same network is sensitive to other high-level concepts such as cat faces and human bodies."[7]

In other words, what Google Brain learned was that it *can* facilitate artificial learning capabilities with limited human guidance. Similar to Ng's optimism, Nicola Jones states that by bringing deep learning closer to what she describes as "true artificial intelligence," Google Brain has enabled a deep exploitation of non-AI tasks, as well as complex computational problems.[8] Jones and other optimists unwittingly overlook several consequences of commercial application. Although portions of Google Brain's algorithms are released to the public under open-source software licenses, the majority of their algorithms are closed source, which means the source code is not published except to licensees. As the owner, engineer, marketer, and distributor of the device, the company determines which data and learning features are most useful and relevant for users and consumers. Moreover, human users and consumers are commercially and computationally valuable only inasmuch as our attributes and behaviors can be exploited for learning and abstracted for classification. This raises significant ethical concerns. Mathew Fuller and Andrew Goffey comment on this confluence: "Being sophisticated today is about operating with media forms, techniques, and technologies that are excessively, absurdly, finalized as to purpose and utility, but whose seductive faces of apparent, personalized seamlessness, whose coded and codified bureaucratic allure, when regarded from the right angle, present multiple occasions (kairos) for crafty—and well-crafted—exploitation, provided that their sleek affection to affectlessness is probed for

7 Quoc V. Le et al., "Building High-Level Features Using Large Scale Unsupervised Learning," *ArXiv:1112.6209 [Cs]*, December 28, 2011, http://arxiv.org/abs/1112.6209.

8 Nicola Jones, "Computer Science: The Learning Machines," *Nature News* 505, no. 7482 (January 9, 2014): 146.

the energy it absorbs."[9] They suggest that algorithmic utility, although promoted as necessary solutions to scientific problems, can articulate nothing more than a tendency to organize the mundane—which can produce less knowledge about the "real world" than a video-sharing service can reveal the mystery behind our obsession with cats.

It is apparent that machine learning and computer science share a similar set of problems, such as the limitations posed by computational efficiency. However, whereas machine learning diverges from computer science in terms of research aims, machine learning and computer science share a commitment to practical knowledge production and distribution. Although the success of these commitments is measured by their accuracy in uncontrolled real-world environments, they lack transparency and reflexivity, which raises additional questions about the dissonances between automated approaches and the utility of scientific knowledge production. Nonetheless, advances in deep-learning techniques have helped accelerate the development of other techniques that benefit from its agility in contingent environments. The development of artificial neural networks is one example. Neural nets are an important device to consider within this context, as they emerge through the mechanistic ideas of brain mapping and cognitive simulation. By the time artificial intelligence reached prominence in the 1950s, machine learning research had already played a central role in the development of AI technologies. Furthermore, with a greater understanding of the structure and functions of the brain—including the neuron and nervous system activity—research within artificial intelligence, psychology, and cognitive science increased at an accelerated rate. In AI, two classes of research emerged that are commonly distinguished today as "traditional" and "connectionism."[10]

9 Matthew Fuller and Andrew Goffey, *Evil Media* (Cambridge, MA: MIT Press, 2012), 18.

10 See Christof Teuscher, *Turing's Connectionism: An Investigation of Neural Network Architectures* (New York: Springer, 2002).

A central principle of connectionism is the description of mental phenomena as a network of interconnected, relational, discrete units. Connectionist approaches were previously known as parallel distributed processing (PDP). Much of this work was done in the 1970s and stressed the parallel relations in nature (specifically mental and behavioral phenomena). PDPs were directly influenced by early work on perceptrons (algorithms for pattern recognition) in the 1950s and 1960s. Perceptron theories by James L. McClelland and David E. Rumelhart demonstrate the robustness of nonlinear and multilayer networks and their viability for a large number of applications.[11]

From a connectionist perspective, neurons in the brain would represent a network of discrete homogeneous units and the synapses would represent the energy transferring connections between them. Connectionist models are mutable and change over time, as any unit in the network can be activated to perform specific functions. An artificial neural network (ANN) is a frequently used connectionist model. Activations are distributed through the probability that a neuron will generate an action potential, or the potential of a specific neuron to excite or depress. There are a number of different ANNs (such as recurrent neural networks, feed-forward neural networks, and common gradient-descent algorithms like back-propagation neural networks), yet they mostly follow two basic principles: that any mental state can be described as a dimensional vector with numeric activation values distributed over the network, and that "memory" is generated by modifying the relational strength (or "weight," a simplified version of the synapse) between neurons. ANN learning is controlled by a set of algorithms and learning rules, many of which are specifically focused on neuronal weight and vector activation. In contemporary artificial neural networks, the activation or sigmoid functions are weighted to simulate the "energy" or information transfer between artificial

11 James L. McClelland and David E. Rumelhart, *Parallel Distributed Processing: Explorations in the Microstructure of Cognition* (Cambridge, MA: MIT Press, 1986), 285.

neurons. Sigmoid functions introduce nonlinearity into the model through pattern classification, clustering, and function approximation.

Cyberneticians Warren McColloch and Walter Pitts were greatly influenced by research on perceptrons. Although they were strongly discredited by prominent artificial intelligence researcher Marvin Minsky, McColloch (a neurophysiologist and psychiatrist) and Pitts (a mathematician) advocated for connectionist approaches to neural systems. Although disputed, Margaret Boden argues that the historical relationship between computational intelligence and representations of mental phenomena find their origins in McColloch and Pitts's pivotal text, "A Logical Calculus of the Ideas Immanent in Nervous Activity."[12]

In the text, McColloch and Pitts argue that the binary quality of neural activity and the relations between neuronal units can be treated as a form of propositional logic. The McCulloch-Pitts thesis finds that the behavior of "nets" is universal, as well as stable in that "for any logical expression satisfying certain conditions, one can find a net behaving in the fashion it describes."[13] The assumption is that logical expressions can be applied to any universal condition in which the behavior of one net comprised of discrete choices is replicated in another net under the same conditions.

Prior to McCulloch and Pitts's search for a logic of the nervous system, sciences of the mind were mere records of observed realities. McCulloch and Pitts, however, adopted what would later become known as connectionist ideas to posit that the structure of the brain, including the signal-distributing nervous system, is comprised of a closed system of elementary components. The neurons within this system follow a binary all-or-none rule of activation. As such, any brain activity—including perception, judgment, and knowledge accumulation and construction (what

12 Warren S. McCulloch and Walter Pitts, "A Logical Calculus of the Ideas Immanent in Nervous Activity," *Bulletin of Mathematical Biophysics* 5, no. 4 (December 1, 1943): 115–33.

13 McColloch and Pitts, 115.

we think of as the mind)—are also functions of binary logics. Importantly, the neural nets were theorized to incorporate an operational memory that loosely mimics that of the human brain. Today, ANNs like Google's Deep Mind differentiable neural computer (DNC) extend this hope by pairing a neural net with a hybrid learning system that learns and forms "memories" by attaching mapping targets to external memory banks.[14]

McCulloch and Pitts's work was profound in its switch from prior research approaches that focused primarily on records of observation to the notion that computation can, with regularity, represent physiological and mental phenomena as operational binaries. Specifically, the logical description of mental atoms (or psychons) and their relation to neuronal pulses gained traction as a new informational theory of science, which would later develop into cybernetic theory.[15] As Mackenzie notes, the ideas renewed "longstanding cybernetic hopes to bringing brains and computation together in models of computational intelligence and agency."[16] The McCulloch-Pitts neural nets were to be important in the formation of several fields of study, including cybernetics, computer science, artificial intelligence, cellular automata, robotics, self-organizing systems, and the trajectory of experimental

14 For more on differentiable neural computers and neural network memory, see Carl Engelking, "An Artificial Neural Network Forms Its Own Memories," *D-Brief* (blog), October 13, 2016, http://blogs.discovermagazine.com/d-brief/2016/10/13/artificial-neural-network-memories/.

15 McCulloch and Pitts's theories resonated with John von Neumann, who in the 1940s designed an artificial computing network for automated evolutionary logic. The aim was to build a self-replicating robot from a vast number of parts. These parts (or cellular automata) would have to be spatially and temporally discreet, homogeneous, self-replicating, and evolve in parallel at discreet stages. The field of study also gained popularity with the idea of universality. It was believed that cellular automata (CA) rules could emulate a universal Turing machine, and therefore compute anything that is computable. Cellular automata are used today for inferential reasoning in neural networks.

16 Mackenzie, *Machine Learners*, 183.

epistemologies that defined logical representations of perception and rules for the dynamic mechanization of mental states.

As nonlinear dynamical systems, cellular automata and neural networks are proficient at modeling specific problems. They also layer informational logics of certainty onto otherwise contingent and observational phenomena, which are thought to be universal, relational with particularity, and repeatable on the level of behavior. As Andrew Ilachinski remarks, discrete models show that "very simple finite dynamical implementations of local conservation laws are capable of exactly reproducing continuum system behavior on the macroscale."[17] Luciana Parisi reminds us: "Rather than demarcating a simple opposition between theoretical and practical knowledge, [...] current computational practices reconfigure the relationship between reason and automation so that automation is no longer a reflection of a Cartesian dualism of mind and body [...] or of the Laplacian mechanization of the universe. One way of theorizing this order of continuity is to address the problem of the limit of computation deductive logic, and formal reasoning as an opportunity to rearticulate the ontology of number in terms of information processes."[18]

Today, ANNs stand at the helm of powerful and ubiquitous research and industry devices. Although ANNs are loosely aimed at simulating the activities of the mind, they also constitute a new mode of machinic perception that weaves the mind (and any pathological representation of it) into the logics of bioepistemic operation. What we witness in the bioepistemic is the captivation of dialectic forms of relation between the natural and artificial, the discreet and collective, or the ordered and chaotic, which remain elemental referents in the formation of computational models. Within these regimes of abstraction and dynamic constriction, life is recast as that which can be collected, traced, and measured.

17 Andrew Ilachinski, *Cellular Automata: A Discrete Universe* (Singapore: World Scientific, 2001), 8.

18 Luciana Parisi, "Automated Thinking and the Limits of Reason," *Cultural Studies ↔ Critical Methodologies* 16, no. 5 (2016): 474.

Machine Learning Models

Modeling is a central function of the machine learning process. A model, in one sense, attempts to capture phenomena by simulating and preempting the organization of partially independent entities in a closed system. Models use statistical techniques such as Bayesian probabilities to identify correlations between system attributes, structures, and visualizations. Modeling begins with an explicit set of assumptions, yet it is often mediated by a combination of mathematical methods, including heuristic intuition and scientific experience. Potential applications for machine learning models are broad and vary depending on the research problem. Mitchell defines machine learning through an example: "A computer program is said to learn from experience E with respect to some class of tasks T and performance measure P, if its performance at tasks T, as measured by P, improves with E."[19] To elaborate: Task (T) might be the desire to classify a tweet that has not yet been published based on the probability it will be retweeted. Experience (E) would be the full canon of tweets on a single user account, some of which have been retweeted and some not. Performance (P) would indicate the accuracy of the classification or the number of tweets on the user account that are correctly predicted to be retweeted.

Machine learning algorithms are optimized to solve these types of problems, which can range from increased computational efficiency to economic, social, or political desires. O'Neil and Schutt argue that the differences between statistical modeling and machine learning reflect cultural differences in the approach to research. For instance, statisticians consider variables in regression models as having "real-world" meaning, defined as being relevant to the research problem. In contrast, computer scientists or artificial intelligence engineers may use machine learning algorithms to interpret parameters for the

19 Mitchell, *Machine Learning*, 3–4.

optimization of cluttering, classification, and prediction.[20] O'Neil and Schutt also note that statisticians are interested in capturing variability or uncertainty by providing confidence intervals and posterior distributions. Machine learning algorithms, like k-means or k-nearest neighbors, share no such notion of confidence and uncertainty.[21] Lastly, O'Neil and Schutt describe the role of statistical models as making explicit assumptions about data processes in learning and which parameters should be eliminated or kept as part of the system, namely through use of probability functions. What O'Neil and Schutt's analysis reveals is the contrast between assessing uncertainty with limited levels of confidence and building models to predict with little or no concern for uncertainty in the data.

However, the distinction between more heuristic types of modeling and computational models may not be as clear as O'Neil and Schutt suggest. Newell and Simon state that "neither machines nor programs are black boxes; they are artifacts that have been designed, both hardware and software, and we can open them up and look inside. We [computer scientists] relate their structure to their behavior and draw many lessons from a single experiment [...] to demonstrate statistically that it has not overcome the combinatorial explosion of search in the way hoped for. Inspection of the program in the light of a few runs reveals the flaws and lets us proceed to the next attempt."[22]

Models rely on consistency, as well as stable operation. Gordon S. Linoff and Michael J. A. Berry refer to the model's

---

20 Schutt and O'Neil, *Doing Data Science*. At its simplest, an algorithm is a set of rules or computational steps to follow to accomplish a task. Parisi has shown, however, that algorithms are not simply instructions to be performed; they are performing entities, or "actualities that select, evaluate, transform, and produce data," that expose the internal inconsistencies of rational systems of governance. Parisi, *Contagious Architecture*, ix.

21 For a deep analysis of k-means and k-nearest neighbor methods, see Latanya Sweeney, "K-Anonymity: A Model for Protecting Privacy," *International Journal on Uncertainty, Fuzziness and Knowledge-Based Systems* 10, no. 5 (2002): 557–70; and Mackenzie, *Machine Learners*.

22 Newell and Simon, "Computer Science as Empirical Inquiry," 106.

ability to successfully preempt structured training data, as well as any unstructured or unknown phenomenon it may encounter.[23] A model's stability is informed by three primary edicts: models are most relevant in "real world" situations that can "effect change of some sort" (the goal of design is not to produce the best possible model in a controlled environment, such as in a laboratory where the data can be overfit); model design should remain sensitive to human programming errors; and models must remain temporally sensitive, since they are influenced by the temporal relationship between input data and forecasted values. Although models are trained on historical data, they may not be applicable for use in other time periods. Linoff and Berry write, "While it is not a requirement that the future be exactly like the past, models work on the assumption that the past is a prologue to the future."[24] As such, the best way to approach temporality is to train the model with input data aggregated as closely as possible to the actual time of model implementation. Otherwise, the model's ability to estimate targets may be contaminated by misaligned variables. For example, input data might represent the exact time a consumer begins a single transaction (or when they first land on the online retailer's website). Time, in these instances, starts from the moment of first entry or arrival until the designated target value is achieved (a customer completes a transaction, leaves the website, returns for another purchase, etc.). A total monetary amount spent, say, over the span of a year has a time frame of one year, starting from the original moment of inquiry or arrival. In this way, the time designation for the input is the same as that for the output: one year—a valuable insight for a marketing agency that wants to assess total spending or advertising exposure over a specified period of time.

23 Gordon S. Linoff and Michael J. A. Berry, *Data Mining Techniques: For Marketing, Sales, and Customer Relationship Management* (Indianapolis, IN: Wiley, 2011).

24 Linoff and Berry, 156.

However, some researchers argue that more incremental temporal procedures can produce greater results for preemption. Temporal methods base performance longitudinally on changes in prediction over time, as opposed to more conventional methods that base learning success on accurate forecasting. Rather, temporal-success predictions or temporal-difference methods take a more incremental approach to assigning actual values to predictions in shorter intervals. The approach is useful in situations where insights with greater frequency are desired. For example, assume a weather forecaster wanted to make a daily prediction that rain might fall on the following Saturday. Using conventional approaches, the weather person would have to compare predictions made on each single day with the actual outcome for Saturday (Monday predicts 50 percent chance of rain on Saturday; Tuesday, 75 percent; and so on). Here, the quality of the preemption is good only inasmuch as rain actually falls on Saturday. With temporal-difference methods, actual results for each day of the week are analyzed incrementally. Data on a particular day is compared not to the forthcoming Saturday, but to days similar to the present day, which are then used to increase the robustness of the overall model. Because predictions are made incrementally, they are easier and faster to compute, and as a result hold an advantage over other supervised learning procedures. The model can be updated daily or often in real time and the algorithm adjusted accordingly. Whereas in conventional methods the question of the weather may be *Will it rain on Saturday?*, with temporal-difference methods forecasters can instead ask *What will the weather be on Saturday?*

Still, temporality is where preemptive models and profiling models differ. While preemption models are organized around the difference between two points in time (input data/forecasted values), profiling models use the same temporal point for both input data and forecasted value. Profiling models rely on static input data, such as demographic information (age, sex, geolocation, etc.), which are then constructed into a profile of the individual. Consumer surveys are a common example. In a consumer survey, unique demographic data is used to construct customer profiles that can then be acted upon. These data remain abstract and dormant in a database until they are desired for use either as future insight for immediate capital gain or historical knowledge for present or future marketing. For instance, insurance companies may use survey data to indicate which premiums are assigned to which customer postal codes. An online retailer, like an airline company, may use profiling information such as internet IP addresses to locate customers and allocate ticket prices according to times of year and customer demand. Even so, preemptive and profiling models are both vulnerable to error. Linoff and Berry write that profiling is limited by its inability to distinguish between cause and effect, not to mention to any common demographic variables that may be hidden within the data. They describe the example of a preemption that suggests men buy more beer than women do.[25] The model assumes certain variables remain fixed while others vary in a unidirectional causality (men drink more beer), instead of alternative correlations that might suggest drinking is instead the cause of maleness. Taken unidirectionally, profiles can conflate social constructions with causality (car theft is higher in neighborhoods with more young Black men, therefore it is likely that young Black men are car thieves). This raises questions about a reliance on the solidity of certain culturally produced phenomena and how they inform an equally fragile causal relation, particularly considering the absurdity of opposing direction (young Black men reside in neighborhoods with high amounts of car theft and therefore it is likely that car thefts create more young Black men).

The tensions illustrate an imbalance in research approaches and operational aims. They, nonetheless, constitute a series of techno-human relations that are fed back into the hierarchal organization of social phenomena. Mackenzie astutely questions what systems are transformed when machine learning is generalized across various epistemic, economic, and institutional settings.[26] But we must question how—in terms of the genesis of Black being—epistemic, capitalist, or institutional settings, or the

25 Linoff and Berry, 158.

26 See Mackenzie, *Machine Learners*, 100.

exclusive inscription of the categorization of being human, are transformed at all in relation to the violences enacted upon the racialized being, or the regulation of Black spatial movements.

Early Learning through Race Play

Machine learning models produce statistics-based knowledges by mapping movement and the empirical relationships between particular entities (also known as agents) in complex environments, such as cities, landscapes, and so on. Models are most often trained in research settings (although recently some models are tested in actual public spaces). The type of agents may be selected at random, but they are typically prescribed depending on the research or industrial target. Agents are not limited to single persons, objects, or entities. They can include homogenous or heterogeneous collectives, clusters, or even environmental phenomena. Agents are represented by description and coded to conduct a number of specific actions. For instance, they may be instructed to make autonomous decisions based on logical deduction, use decision-making to resemble practical reasoning in humans, combine deductive and other decisional frameworks, or explicitly not use reasoning at all. These decisions can be made in either controlled environments or environments with various levels of contingency.

Agent-based models (ABMs) are one type of approach to targeting based on agent maneuverability. ABMs simulate and predict the behavioral patterns of semiautonomous agents, which may learn, adapt, or reproduce within the parameters of a predetermined rule set. Variable selection can also be determined based on historical agent performance. In other words, ABMs can use error to strengthen and systematically reveal what independent behaviors might correlate with larger patterned behavior. Thomas Schelling's "Models of Segregation" from 1969 is one of the first functional ABMs—as well as one of the field's most controversial. Schelling designed the model to gain insight into the dynamics of racial segregation in urban areas in US cities. Instead of using arbitrary parameters to establish his rule set,

Schelling used a normalized set of classifications—sex, age, income, first language, color, taste, comparative advantages, and geographical location—which were assumed to be equitable relations among each semiautonomous agent. The system distributed equally Black (B) and white (W) agents across the spatial grid. Once their starting points and attributes were randomly established, the agents were instructed to make a decision on the optimal place to move given their relationship to their nearest neighbor and the general surroundings, which Schelling classified under the parameters "Contentment" and "Neighborhood." The Black and white agents were governed by a simple instruction: if the agent in a neighborhood with a radius of one agent is content with a 50 percent mix between it and other Black and white agents adjacent to it, then the agent can remain stationary; if the agent exceeded the contentment threshold, then it was allowed to move until it reached contentment.

The results of Schelling's experiment suggest that there are thresholds of contentment that justify self-segregation in heterogenous environments. For instance, a line of agents BBBWBWWW in a neighborhood of radius one will rearrange themselves to become BWBWBWBW after several iterations. However, if the neighborhood extends to two of the same type of agents (BBWBWBBW), the outnumbered agent will surpass its contentment threshold and rearrange. In other words, individual white agents, for instance, will surpass their contentment thresholds when there are many more Black agents in their immediate surroundings. Alternately, Black agents will surpass their contentment thresholds when there are many more white agents nearby.[27]

In more recent computational interpretations of Schelling's model, algorithms determine whether or not agents are content or discontent with their locations. If an agent is not content, the algorithm selects a new home or adjacent square for the agent. If the square is empty then the agent will move to this location.

27 Thomas Schelling, *The Strategy of Conflict* (Cambridge, MA: Harvard University Press, 1960).

If not, the agent is forced to stay in place. The learning exercise adds relevancy to Matthew Fuller and Graham Harwood's critique of reductive data practices. They write:

> The specific categories upon and through which segregation operates are described as if natural, not even worthy of equivocation as to their relation to social structure. The racism of the work is both that it operates by means of racial demarcation as an autocatalytic ideological given and secondly that it provides a means of organizing racial division at a higher level of abstraction. To say that Schelling operates within an ideologically racialized frame is not to aver either way as to whether Schelling as a person is consciously racist, but that, in these papers, racial division is an uncontested, "obvious" social phenomena that can be reduced in terms of its operation to a precise set of identifiers and operations.[28]

Fuller and Harwood point to the subtlety of essentialist narratives that dislocate human accountability in the abstraction of racialized computational logics and the invention of phenomenal justifications. As Stuart Hall asks, "Do we bear our cultural identities, the signs and symbols of our 'belongingness,' like (as some would argue) indelible number-plates on our backs? Or are the unities that cultural identities appear to constitute the result of what we might call a 'practice of narration,' the invention of the cultural self, producing fixed belongingness in rather the way we construct, after the event, a pervasive, consistent biographical 'story' about who we are and where we came from?"[29]

Arthur Samuel's Checkers-Playing Program from 1959 is one example. Considered the first self-learning model, Checkers-Playing Program has greatly contributed to the early advancement of artificial intelligence and machine learning research. The model evaluates the dynamic spatial positions of individual

28 Matthew Fuller and Graham Harwood, "Abstract Urbanism," in *Code and the City*, ed. Rob Kitchin and Sung-Yueh Perng (Abingdon: Routledge, 2016), 63.

29 Stuart Hall, "Forward," in Claire E. Alexander, *The Art of Being Black: The Creation of Black British Youth Identities* (Oxford: Clarendon Press, 1996), v.

entities on a designated grid, like pieces on a checkers board. The pieces, however, are semiautonomous and governed by a set of parameters that are dynamically modified with each move, depending on each piece's original starting position and their relation to other pieces on the grid. In other words, each piece moves based on its own parameters, which are also determined by the relation the pieces have with their surroundings. The program simulates both the movement and anticipation of relations within a closed space, where final outcomes are predicated on the unfolding of smaller discrete interactions. This "learning procedure," as Samuel calls it, is frequentative.[30] The model is further optimized with each iteration. In this way, any historical errors are beneficial to the final outcomes, as they provide specific insights into how the overall parameters can be adjusted to reach the original hypothesis. As Richard S. Sutton and others note, the success of Samuel's program was in its capacity to self-learn to the point where it gained enough knowledge of the overall system to play against a human being.[31]

Similar to Samuel's Checkers-Playing Program is John Holland's Bucket Brigade from 1985, an adaptive system called a classifier system, in which conflict resolution is carried out by competitive action. Specific rules in Bucket Brigade are dynamically adjusted with each iteration based on the movement of semiautonomous entities within the overall system.[32] They depend less on the final outcome and more so on the relation between the historicity of movement and the potential for future interaction. Unlike Checkers-Playing Program, however, rule sets within in Bucket Brigade are modified based on numerical parameters. The rule sets are preassigned unique messages that

---

30 Arthur L. Samuel, "Some Studies in Machine Learning Using the Game of Checkers. I," in *Computer Games I*, ed. David N. L. Levy (New York: Springer, 1988), 335–65.

31 Richard S. Sutton, "Learning to Predict by the Methods of Temporal Differences," *Machine Learning* 3, no. 1 (August 1, 1988): 9–44.

32 J. H. Holland, "Properties of the Bucket Brigade," *Proceedings of the 1st International Conference on Genetic Algorithms* (July 1985): 1–7.

are matched to a list. If a rule successfully "bids" or matches its message with one on the list, it gains "strength" and is awarded the right to post its message to the working memory data structure. Within this competitive structure, the rule set with the highest number of successful bids becomes active and is allowed to post its message to a new list, which is then used in the next round of the game. An active rule will lose strength, however, unless it can associate itself with other rules.

What's most significant about Bucket Brigade is that it gives a competitive advantage to certain rule sets over others. The simple interplay of rule exchange builds a chain of rule invocations that are learned through an exchange of numerical insights, which are passed back and forth from one rule set to another. As such, only the strongest rules survive. Today, traces of strength-based learning are found in search algorithms that prioritize certain webpages based on popularity or current rankings, as well as user preference based on other pages. Strength can also determine recommendation functions that display outcomes based on prior user interests and the strength of other users' associative searches and purchases.

Support vector machines (SVM) are more advanced forms of modeling. A specific type of a supervised learning model, SVM are designed to analyze data using various machine learning techniques such as classification, regression, and pattern recognition. One of the significant characteristics of SVM research is its shift away from previous emphasis on probabilities. Instead, it focuses on representation and entity positioning within abstract space. Cristianini and Shawe-Taylor describe SVM as "learning systems that use a hypothesis space of linear functions in high dimensional feature space, trained with a learning algorithm from optimization theory that implements a learning bias derived from statistical learning theory."[33] These representations are useful for visual graphics, since present input data can be mapped onto historical models to preempt future

33 Cristianini and Shawe-Taylor, *An Introduction to Support Vector Machines*, 7.

data trajectories, including any alignments or deviations from baseline behavior.

Predetermined structure or pre-categorization are not necessary for classification with SVM. Classification can be preempted according to weighted similarities between attributes based on any variability between these attributes and baseline normalities. In other words, entity classification is determined by attribute similarity, regardless of any dissimilarities in other attributes. Ultimately, SVM learn through association and commonality outside of explicit historicity—as a young child might. For instance, a newspaper article might be classified as similar to another if they share similar words or the same writing styles. A sports car might be distinguished from a family sedan based on commonly shared attributes such as "sporty" aesthetics, quick acceleration, agility, and so on. If the family sedan, such as a Porsche Cayenne, shares these attributes it can be classified as a sports car. If not, it may be classified into another segment.

SVM can also cluster individuals based on abstract descriptions. For instance, an individual may be classified as a potential threat if their gait is aligned with behavioral markers that indicate stress, erratic behavior, and velocity, which in terms of behavior psychology may indicate a person highly distressed and therefore suspicious. Following this example, SVM are used in various surveillance applications, ranging from biometric emotion-detection systems to facial recognition. As such, SVM can preempt future behavior, as well as distinguish between behaviors that align with baseline normalities and those that deviate from the established range. Like ABM, SVM can evolve through dynamic learning processes that enable them to update their historical information and preemptions in real time. This is significant considering the role modeling played in the early developments of machine learning.

The logics of modeling force feed any variability in agent interrelation into empirical assumption that are further regressed into probable facts and assurances. The grouping of certain peoples around controlled substances—in this instance, race—is troubling, as it exists as an exercise of sociogeny. It is prototypically devoid of indeterminacy and historical (and present) cultural organizations around racial hierarchies, economic and spatial weaponization, and colonial intent. Data and the phenomenon of being—as being-in-relation to race—are simplified in the service of eugenic determinacy, further naturalizing economic struggles for mobility, racial aggressions, and self-segregations as a logical form of pleasure and happiness.

Nonetheless, there is a profound power in the act of modeling. Some modeling practices reinforce the perceived divides between the universal substances of race and discursive understandings of particular racialized operations. It also further detaches the genesis of inequitable relation from the production of the machinic. The full birth of the social is irreducible to mathematical approximation. However, the limits of mathematics are its greatest asset when the administration of power relies on universalism as a prototypical model and the exclusion and segregation of the Other to justify the creation of "a patchwork of overlapping and incomplete rights to rule."[34] This is done, as McClintock argues, to invent race in the urban metropoles as the very self-definition of the middle class as well as the very necessity for the policing of the "dangerous classes: the working class, the Irish, Jews, prostitutes, feminists, gays and lesbians, criminals, the militant crowd, and so on."[35]

Some models have been widely criticized for reinforcing racialized perception. One of the most controversial examples is a 1959 study conducted by criminologists Sheldon and Eleanor Glueck. In *Predicting Delinquency and Crime*, the Gluecks used variables such as age, IQ, and neighborhood of residence to determine any possible correlations with behavioral patterns, like family stability, affection, and discipline in the home.[36] The Gluecks concluded that based on these factors, some juveniles were more likely to become socially delinquent than others. Unsurprisingly, the study was met with considerable objection.

34 Mbembe, "Necropolitics," 31.
35 McClintock, *Imperial Leather*, 5.
36 Sheldon Glueck and Eleanor Glueck, *Predicting Delinquency and Crime* (Cambridge, MA: Harvard University Press, 1959).

Critics argue that the model ignores the social, political, or economic conditions that might contribute to the correlations, and that it assumes an objectivity where none exists. The Gluecks insisted that their model was not meant to determine delinquency with any certainty, but instead provide only probable insights based on the above factors. They also argued that although prediction tables should be flexible enough to accommodate the dynamic conditions of life, models are not designed to account for every possible variable influence. As Harcourt argues, random sampling is a tangible alternative to policing, as it blinds justice to anything other than objective forms of preemption in the form of potential criminality.[37] Like the Gluecks, Harcourt's assertion reinforces the illusion of objectivity as the condition for justice, and lessens the role of institutional racism and empirical justification play in the circulation and recirculation racialized subjects based on the similarity of characteristics.

If we are to recognize ourselves as subjects of machine learning then we might consider how the fictive substances of race and racialization might be, for some beings, the localized individuation of the power-knowledge relation, a nuance left behind in Harcourt's universal assessment. The provocation, however, isolates the composition of racial substance as that which exists outside of universalization and is rather the mundane ritual of racialized violence. Mackenzie's text is an important intervention, as it makes apparent that the type of world-making that machine learning algorithms are coded to enact are not limited to race or difference; instead they are deeply embedded in the production of thought. Although the mutable layers of power-knowledge are assembled in machine learning practice, they are underwritten by the preemptive function of variation and probability distribution.

37 Bernard E. Harcourt, *Against Prediction: Profiling, Policing, and Punishing in the Actuarial Age* (Chicago: University of Chicago Press, 2006).

Distributed Probabilities as the "Guide to Life"

Mackenzie writes: "Machine learning reverse-engineers the invention of modern statistical thinking. It also takes back the 'real quantities'—probabilities—that modern statistics had attributed to the populations in the world and distributes them to devices, to machine learners that people then observe, monitor, and indeed measure again in many ways. The direct swapping between uncertainty in measurement and variation in real attributes that statistics achieved now finds itself rerouted and intensified as machine learners measure the errors, the bias, and the variance of devices. Although it relies heavily on probability distribution, machine learning is a fat-tailed distribution of probability."[38]

In his groundbreaking book *The Taming of Chance*, Ian Hacking describes how theories of probability are thought to have originated with Blaise Pascal and Pierre de Fermat (in studies on gambling risk), but it was Pierre-Simon Laplace who extended the theories to account for a universal base of reason. In his 1795 text *Philosophical Essay on Probabilities*, Laplace introduces a formal rule of succession based on the contingency of jury decisions. The proposition incorporates historical data on historical juridical decisions to identify the probability that certain types of material evidences might or might not influence human perception and decision in the assignment of innocence or guilt. Laplace's theories greatly contributed to the construction of preemptive knowledge aimed at articulating a universal and achromatic symbolism of movement through the legal system.[39] Laplace writes: "Given for one instant an intelligence which could comprehend all the forces by which nature is animated and the respective situation of the beings who compose it—an intelligence sufficiently vast to submit these data to

38 Mackenzie, *Machine Learners*, 106.

39 Pierre-Simon Laplace, *A Philosophical Essay on Probabilities*, trans. Frederick Wilson Truscott and Frederick Lincoln Emory (New York: JCosimo, 2007), 4.

analysis—it would embrace in the same formula the movements of the greatest bodies of the universe and those of the lightest atom; for it, nothing would be uncertain and the future, as the past, would be present to its eyes."[40]

The early workings of the bioepistemic claim a de facto independence from the complex links between bodies that are shaped by the criminal justice system and those that shape the system itself. The technologies of decision-making form a tactile obedience to the algorithmic process as the zone of exception, where power is turned away from discipline and placed into the mechanization of judicial visions of order and compliance. This mathematical "guide of life," as Joseph Butler remarks, limits levels of complexity under the logics of divine objectivity.[41] Nonetheless, the reach of Laplace's probability theory is augmented by modern mathematical processes—for instance, Bayesian reasoning, which is a simple mathematical formula for the reduction of complex variation into symbolic representations of probable truths. Bayesian probabilities continually adjust variable estimates on the basis of dynamic observational assumptions by simplifying data into more manageable variables. Bayesian probabilities are particularly attractive tools for use in data-driven research, since they enhance computational speed while also optimizing algorithmic power. Today, Bayesian reasoning is an essential tool in machine learning research, which operates in highly complex and contingent environments.

Some researchers argue that problems arise when probabilities are viewed as objective and neutral intelligences, particularly when they are inferred from prior knowledges and hypotheses, which are driven by assumptions of impact and causality. Researchers contend that these assumptions carry no standard practice and can be infused with subjective biases which may influence important decisions, as well as the field of

40 Pierre-Simon Laplace, quoted in Ian Hacking, *The Taming of Chance* (Cambridge: Cambridge University Press, 1990), 12.

41 Joseph Butler, *The Whole Works of Joseph Butler* (London: William Tegg and Co., 2016), 64.

vision from which these decisions are derived. Oscar H. Gandy suggests that "assumptions about prior probabilities are made by members of juries, as well as by all decision makers in the complex decision chain that begins with observation or surveillance on the street. Assumptions include those made by the police officer whose decision to stop the car was influenced by his beliefs about black drivers, as well as the prosecutor who decides to seek a felony conviction based on his beliefs regarding the social impact of crack cocaine. Their decisions are all influenced by their prior probability estimates."[42]

Others, such as Adnan Darwiche, argue that while the chain of observation may appear to suggest that degrees of belief are held primarily within the subjective, probability is that which instead governs the construction of the subjective, as well as any perceived objectivity.[43] This impasse also reveals important concerns regarding the extension of symbolic interpretation to phenomena that cannot be materially observed. All probabilities, according to Darwiche, are degrees of belief that reflect a state of knowledge of the particular, and do not necessarily correspond to anything that can be measured by a physical experiment.

In machine learning, probabilities raise additional concerns about scale. Large-scale applications can consist of hundreds or thousands of variable inputs, each holding their own margins of error. Stacking these errors risks extending probabilistic determinations beyond what is justifiable. Nonetheless, Yaser S. Abu-Mostafa, Malik Magdon-Ismail, and Hsuan-Tien Lin argue that a probabilistic view can produce satisfactory results without assumptions outside of those produced independent of the hypothesis.[44] Advocates assert that, in many cases, experts are trained to intuit forms of uncertainty. They

42 Oscar H. Gandy, *Coming to Terms with Chance: Engaging Rational Discrimination and Cumulative Disadvantage* (Abingdon: Routledge, 2016), 71.
43 Adnan Darwiche, *Modeling and Reasoning with Bayesian Networks* (Cambridge: Cambridge University Press, 2009).
44 Yaser S. Abu-Mostafa, Malik Magdon-Ismail, and Hsuan-Tien Lin, *Learning from Data* (AML, 2012).

insist that as long as engineers use the same distributions consistently for each problem set in each stage of learning, any prior knowledges are unnecessary. They claim that debates on the subjective are misaligned with the aims of probabilistic learning, as probabilities are not expected to replicate target functions perfectly from their origin. Instead, probabilities are meant to approximate correlation in controlled environments, with an awareness that performance outside of the laboratory may vary. To mitigate any discrepancies, measures of error can be used to compensate for unknown variation. It is further argued that while Bayesian probabilities are premised on the selection of only some variables as material evidence, any adjustments necessary to compensate for error provide new evidence in support of new, more informed hypotheses. Humean inductive reasoning prioritizes the number of observable instances in establishing a relationship with the production of knowledge. For Hume, scientific judgment is based on the probability of observable outcomes: the more instances, the more probable the predicted conclusion. Critics assert that the fragility of these types of Humean hypotheses originate in the priority they place on scientific judgment. Michael Wood argues that without a more complete understanding of the role of the subjective within the determination of probabilities, they remain assessments of ignorance and judgment. He states: "If, for practical reasons, samples are not selected randomly, the question then arises of whether they can reasonably be regarded as if they were selected randomly. This is a matter of judgment."[45] The matter of judgment is what Gandy sees as the fundamental determinant of subject position. "How we evaluate people, places, and things in terms of their departure from what we have defined as the norm," Gandy states, "is often a fundamental determinant of the position they will come to occupy in still other distributions that we have yet to consider."[46]

45 Michael Wood, *Making Sense of Statistics: A Non-mathematical Approach* (Basingstoke: Palgrave, 2004), 24.

46 Gandy, *Coming to Terms with Chance*, 4.

Gandy is skeptical of mathematical applications that achieve a special status in the production of knowledge, especially when these applications are used to determine subject positions based on chance, risk, and preemption. Gandy asserts that the use of mathematical tools has foreclosed other traditional ways of revealing social problems and balancing competing interests—more so when these tools become "essential component[s] of any claim of expertise within the most professional domains."[47] Gandy describes this foreclosure as rational discrimination, or the sites where social inequity is determined by knowledges derived from empirical proofs. As Gandy suggests, we have come to trust these positions as authorities over other forms of intelligence, judgment, and narrative. Rational discrimination can be further conceptualized as a site of desire, or an encounter that provides desire with a means to express itself as a site of prototypicality. Although probability-theory techniques are based primarily on estimates, they are thought to be viable alternatives to human knowledge. The act of categorization presupposes a correlation between the significance of certain humans and a set of racial codes and standards.

As Brandom reminds us, "Description is [merely] classification with consequences, either immediately practical ('to be discarded/examined/kept') or for further classification."[48] To dislodge sentient performances of normality, the classification might be inverted to describe entities not as actual, but as an iterative form of individuation. Here, the rule set is operative as an ordering function only inasmuch as it is self-aware of its own condition for the genesis of new emergent relations. From a machine learning perspective, the task of science and ontology would align with a relational process that, although in exchange with the biological and a priori classification, is open to the potential for nonhuman forms of reason.

---

47 Gandy, 8.

48 Robert Brandom, "How Analytic Philosophy Has Failed Cognitive Science," in *Proceedings of the Workshop on Bob Brandom's Recent Philosophy of Language: Towards an Analytic Pragmatism* (Genoa: TAP, 2009), http://ceur-ws.org/Vol-444/paper13.pdf.

CHAPTER 5
Where All Things Are Bound by Race and Data

> I say that intelligence has never saved anyone; and that is true, for, if philosophy and intelligence are invoked to proclaim the equality of men, they have also been employed to justify the extermination of men.
>
> —Frantz Fanon

Whether machine learning's data-rich variants direct us toward a seemingly targeted goal of racial sorting or whether they serve as more subtle points of reference to a specific identifiable position in the continuum of racial difference, algorithms are marked by the practical demonstration of perceptual distinction and acute forms of governance. An erroneous conclusion to this drama is characterized by the act of improving or expanding upon the influence of whiteness as the purveyor of activities, supplying or providing intelligence for a future that contains the known liability of Blackness toward the provision of authority. The potency of this act is entwined with other historical models of human classification, namely in seventeenth-century Europe. Although the aggregation and statistical analysis of data did not originate in Europe, or in the seventeenth century, the period marks an important shift in the collective use of data throughout European countries. Prior to the seventeenth century, there were very few mathematical methods to explain the world. However, as new statistical methods began to emerge, the world and the people within it came into view as an exercise in statistical correlation.

Much of the statistical activities during the seventeenth century were championed by amateur statisticians who had a personal interest in social observation.[1] It was believed that new sources of data could offer clues to help statisticians better understand the world around them. Amateurs formalized their observations into small "life tables" (also known as "mortality tables" or "actuarial tables"), which contained semi-anonymous data on individual characteristics such as age, place of birth, trade, "fighting men," the number of burials, and, in many cases,

1 See Hacking, *The Taming of Chance*.

cause of death. Markers such as these were used to make quantitative judgments by seeking numerical regularities within the data, which could be attributed to what was thought to be a causal relationship between human social patterns and the laws of nature. Amateur hobby coincided with more a general public fascination with scientific advances in the areas of astronomy, geometry, optics, music, and mechanics. Among these discoveries was a new statistical understanding of human life as that which can be approximated with the causal certainty of natural phenomena such as the consistent rising and falling of the sun or the growth of spring harvest.

As data aggregation increased in popularity, statistics became the preferred method for empirically driven knowledge production. Life tables expanded into myriad subclassifications, such as the rate of urban versus rural death, the proportion of tradesmen afflicted by a certain chronic disease, or the excess of male compared to female births. Data and statistics became powerful tools for human classification and the serious study of human-impacted phenomena such the growth of domestic commerce and international trade, the reproduction of labor forces, and the impact of epidemics on political economies. Prior to the question of the bioepistemic and the naturalization of sociogeny, the conditions for ethnophenotypic classification were already set forth by early statistical methods, which are in close kinship to machine learning.

Algebraici et infinitesimalis

The etymology of the word "statistics" is illustrative, particularly considering that it derives from the same root as the word for "state" from the Latin *status*, meaning "manner of standing, position, condition."[2] This reflects the origin of statistics as a

2 The tradition of "university statistics" can be traced to German philosopher, economist, and statistician Gottfried Achenwall in his *Staatsverfassung der Europäischen Reiche im Grundrisse*

process of data manipulation already subsumed in the logics of administration. According to Hacking, "every state [...] was statistical in its own way."[3] So much so that by the late seventeenth century, European colonial nations had already begun collecting survey and census data on their respective colonies to aide in the facilitation of local resource extraction.

Although some colonial territories administered their own data processes to evaluate resource metrics, for colonial nations the physical body as resource and labor was as crucial to the optimization of colonial capital expansion as nonhuman material properties. More so, colonial capital expansion relied upon the universalization of these processes to optimize resource production. Colonial expansion was supported by the work of empiricists who believed that data could provide direct experiences and engagements with distant territories, as well as a universal language of enumeration to help engineer and facilitate logistics. The maximization of capital depended upon the ethnophenotypic sorting of colonized bodies.

The theories of Leibniz, Crombie, and others, whose work systematically correlates biological phenomena with mathematical logics and patterning, assisted, if even implicitly, in enacting this use of data. Take for instance, Leibniz's prolific summaries of data, as well as his differential calculus for the study of applied mathematics. His law of continuity and transcendental law of homogeneity, stated in his 1710 text entitled *Symbolismus memorabilis calculi algebraici et infinitesimalis in comparatione potentiarum et differentiarum, et de lege homogeneorum transcendentali*, established a heuristic, or intuition-based, approach to

(Constitution of the Present Leading European States, 1752). Achenwall is widely credited as the founder of statistics or *statisticum*. Others, however, claim that although we owe the word "statistics" to Achenwell, the techniques were based on the findings of seventeenth-century British economist William Petty, with his use of mathematical methods in political-economic analysis, also known as political arithmetic. See Wood, *Making Sense of Statistics*; and Hacking, *The Taming of Chance*.

3 Hacking, 16.

the study of human behavior and quantifiable universality.[4] In the text, Leibniz formulates his theory of universality that he later proposed as a prototype for a central branch of government. The aim was to provide governments with tools to estimate population phenomena, which at the time was limited to the aggregation of basic biometric data such as time of birth and death, cause of death, gender, and so on. Leibniz envisioned a government office that was flexible enough to serve various interests of the state, including military, civil, mining, forestry, policing, and other data.

The quantification of population phenomena is a key factor in Leibniz's work. After all, Leibniz's proposal to the state was predicated on his belief that the true measure of state power resided in its population—not as a mechanism of democratic agency, but in the form of the quantification of this agency for state control. Under this rubric, Leibniz conceived that the measurement of state prosperity, and thus the maintenance and expansion of power, was measurable by knowledge as a problem of data. In other words, in order for a state to prosper it must produce and maintain a continual stream of insight on the behaviors of both human and nonhuman resources. This knowledge had to be flexible enough to account for the particulars of contingent phenomena—for example, surveys of the number of men in the population to forecast potential military power, harvest forecasting to predict military food supplies, and other observable phenomena. The continuation of these theories is no better illustrated than in contemporary practices that assign birth and death records to national numbers, which are centralized and correlated with other government interests, like

4 Judea Pearl defines heuristics informally as a process "popularly known as rules of thumb, educated guesses, intuitive judgments, or simply *common sense*." More formally, Pearl elaborates on heuristic learning as "strategies using readily accessible though loosely applicable information to control problem-solving processes in human beings and machine." Judea Pearl, *Heuristics: Intelligent Search Strategies for Computer Problem Solving* (Reading, MA: Addison-Wesley, 1984), vii.

taxation, insurance, and consumer records. These data create national profiles that, unlike earlier practices, do not require the individual to be present in order to be accounted for. Instead, statistical citizenship leaves trails of behaviors aggregated into tables (today, databases) for future use in forecasting relationships between mobility and state-determined best practices.

The importance of Leibniz's concept is in its capacity to statistically associate biological phenomena with capital and state value. Fundamental to Leibniz's proposal is the ability to observe as well as act on any insights derived. In this way, Leibniz's theory can be thought as an early mode of data-driven population governance. While these concepts were not actualized until much later, Leibniz's interest in enumerative logics eventually materialized into early census and classification tables. By 1730, Leibniz's classification tables gave rise to more official, administratively sanctioned data science. As Hacking suggests, although it originated in the practices of amateurs, data emerged as a political strategy. Data informed ideas of optimization and the stabilization of contingency as much as it did secrecy and forms of power and surveillance. For instance, Leibniz believed that if populations can be represented as measurements of state power, then similar observations can be made by the enemies of a government as well. Thus, data aggregated by the state needed to be secured and protected. Here, the potential for reconnaissance equated the abstraction of data with proprietary insight and state secrets, enacting an early form of black-box military processes. As such, data was used to stabilize social systems as well as administrative and military processes.[5]

5 Also at stake is the ease with which moral and political economy have maneuvered between lines of mathematical logics. Consider the Prussian military general and strategist Carl von Clausewitz's rational calculus of war termination: "Since war is not an act of senseless passion but is controlled by its political object, the value of this object must determine the sacrifices to be made for it in magnitude and also in duration. Once the expenditure of effort exceeds the value of the political object, the object must be renounced and peace must follow." What is particularly interesting about Clausewitz's theory is not so much its provocation of a

Difference and the Divine

By the eighteenth century, population surveys had become a phenomenon of their own. Not only had they risen in popularity within governance, but they were published widely in private and commercial sectors as well. It was a commonly held belief that natural phenomena, like sickness and death, were contingent rather than preemptive clauses; nonetheless, statistics were still understood as a science that could enable the estimation and forecasting of these observable processes. Alain Desrosières

---

moral means to war, but the scale at which war is carried out, not in strength or even violence, but the calculation of political will. For Clausewitz, war is no more than the larger magnitude of a duel, a one-to-one confrontation, where the immediate aim is to incapacitate an opponent in order to prevent any resistance. In other words, war is a direct confrontation between two discreet elements, the victor of which is the one able to use the relationship between their differences to invoke submission, compliance or surrender. Clausewitz argues that "countless duels go to make up war," but the operation of war is more than the magnitude of force of one aggregate against another of equal, less than, or greater sum. The calculation of war exceeds the relation between the immediacy of incapacitation and the magnitude of force required to do so. Instead, war is "an act of force to compel our enemy to do our will." Carl von Clausewitz, *On War* (Princeton, NJ: Princeton University Press, 1976), 92.

Here, force is a self-imposed, imperceptible imitation of the self that voluntarily submits to the will of the other. Clausewitz states: "If the enemy is to be coerced you must put him in a situation that is even more unpleasant than the sacrifice you call on him to make. The hardships of that situation must not of course be merely transient—at least not in appearance. Otherwise the enemy would not give in but would wait for things to improve. Any change that might be brought about by continuing hostilities must then, at least in theory, be of a kind to bring the enemy still greater disadvantages. The worst of all conditions in which a belligerent can find himself is to be utterly defenseless. Consequently, if you are to force the enemy, by making war on him, to do your bidding, you must either make him literally defenseless or at least put him in a position that makes this danger probable. It follows, then, that to overcome the enemy, or disarm him—call it what you will—must always be the aim of warfare." Von Clausewitz, 77.

argues that keeping records on baptisms, marriages, burials, and other rituals and life phenomena is linked to concerns with determining individual identity, thus existence, as a state subject for judicial and administrative means.[6] The "explosion" in data aggregation, as Hacking describes it, multiplied exponentially into a myriad of subclassifications of both living and nonliving phenomena, such as animal livestock and commercial infrastructure.[7]

One of the most significant of these types of study was conducted by German demographer Johann Peter Süßmilch. Like Leibniz, Süßmilch used parish records to investigate gendered birth occurrences. In *The Divine Order in the Circumstances of the Human Sex, Birth, Death and Reproduction* from 1741, Süßmilch outlines the discovery of what he believed to be a linear ratio between male and female birth rates. Although natural childbirth was generally thought to be a contingent phenomenon, Süßmilch argued that childbirth was instead the work of the divine. As such, any patterns found within the process of birth could be reflected by a mathematical representation. Using these logics, Süßmilch concluded that males were not only born at a higher rate than females, but the number of future births could be forecast as divine will. These data were then localized to correlate birth rates among various geographic locations. Süßmilch correlated these data into a general principle of morality that reduced the contingency of childbirth into a theory of divine provenance. Cities and territories with high birth rates were seen as more favorable by God, and therefore of higher social order than populations with lower birth rates. Süßmilch's principle was reinforced by earlier postulates by Scottish mathematician and physician John Arbuthnot, who, in 1710, uncovered similar regularities in male and female birth rates. As with Süßmilch, Arbuthnot attributed his correlatives to provenance.

6 Alain Desrosières, *The Politics of Large Numbers: A History of Statistical Reasoning*, trans. Camille Naish (Cambridge, MA: Harvard University Press, 1998), 23.

7 Hacking, *The Taming of Chance.*

Hacking argues that these correlatives reflect early tendencies to conflate data with justifications for the establishment of social hierarchy—a convention that Immanuel Kant would later adopt to elaborate on his own theories of birth and death as justifiable "conformit[ies] to the laws of nature."[8]

Race as Natural Law

Data analysis was also enacted to justify the construction of race as scientific fact. The first tables on immigration, emigration, nationality, and race were created in 1745, after the publication of *The Divine Order*. The tables aligned with anthological travelogues based on empirical observation of indigenous populations. Aside from tables on birth, death, and mortality, data was also recorded on racialized physiological, perceptual, cognitive markers, such as skin color, image recognition, and skeletal measurements. By 1900, an enthusiasm for race classification exploded into methodologies in philosophy and the human sciences aimed at demonstrating racial hierarchy. In many cases, biological data was correlated with patterns found within natural phenomena. Data also strengthened Anglo-European ideas of racial superiority as a natural, and thus divine, process of evolution. Tukufu Zuberi argues that one of the most provocative examples of racial classification is the Great Chain of Being, a strict medieval catalogue of hierarchal structures of all matter and life thought to be decreed by God. The chain uses a Platonic and the Aristotelian concept of *historia animalium* to vertically stack the material universe, starting with God as the perfect being and descending downward to angels, demons, astrological matter, kings and queens, and on to commoners, animals, plant life, and so on. The "links" in the chain were subdivided into further categories, with humanity as an intermediary between spiritual beings and physical objects. Based on this premise, humans were thought to possess powers of

8 Hacking, 18.

reason and imagination similar to the divine. Within this sorting of knowledge, humans were further sorted by wealth and social status. For instance, those with the highest status, such as kings and queens, were ranked higher than, say, servants or slaves. Therefore, kings and queens were closer to the divine than other humans. As such, those on the lower end of the spectrum, such as the servant or slave, were ranked just above the "highest" ranking beast, which was the ape. Humans were also thought to possess other attributes, such as sensation, which in the divine sense gave humans access to all things knowable. Nonetheless, humans are still bound by the physical forms of the body and organs, which under the schematic created a tension between divine forms of knowledge and what was thought to be the human's lower animalistic nature. With the divine order in place, any sorting of lower-ranked humans, such as Africans or Hottentots—including their segregation, extraction, displacement, ownership, or disposal—could be justified as both natural and mathematically sanctioned. In addition, colonial administrators and publics could be relinquished from any direct accountability for their action. Paradoxically, as quasi-humans, Hottentots were thought to still maintain a sensory and imaginative view of the world. However, their rationality and reason were decreed as limited by what was believed to be their predominant animalistic nature.

Racial hierarchies were exacerbated by eighteenth-century developments in evolutionary theory in biology and social statistics. As Zuberi writes: "When the Great Chain of Being no longer carried the weight of legitimacy, science came to the rescue."[9] Zuberi makes reference to Swedish natural historian Carl Linnaeus, who was one of the first scientists to formulate an empirical order within human populations. Linnaeus's tables, published in 1735 in *Systema Naturae*, correlated physiological data on human populations with animal species. Although Linnaeus did not explicitly rank species in a hierarchal scale,

9 Tukufu Zuberi, *Thicker than Blood: How Racial Statistics Lie* (Minneapolis: University of Minnesota Press, 2001), 19.

he did suggest, aesthetically, that Hottentots did not derive from the same origin of species as Anglo-European people. Linnaeus's charts had a great influence on research in evolutionary theory. For instance, Jean-Baptiste de Monet de Lamarck adopted Linnaeus's chart to develop a theory of evolution based on heredity or the inheritance of acquired characteristics, dismissing the role of random biological variation. Lamarck's study had an impact of its own, particularly through the work of Thomas R. Malthus, who, in 1798, applied characteristics of survival to population growth phenomena. Malthus argued that population growth can be controlled strictly on the basis of statistical evidence. He further argued that population growth was a geometric phenomenon, while resources, such as food, grew arithmetically. Malthus's theory was consequential to public and administrative views on poverty and homelessness. Malthus drew upon earlier work in evolutionary theory to reinforce the naturalization of the poor as a postulate of social patterning and, thus, divine provenance. With the poor relegated to a lower category of human—at a greater distance from God than the wealthy—then any attempt to aid the poor under the postulate would be a rational act against God.[10]

It was in this context that medical statistician William Farr articulated a close relationship between national, tribal, and family group. For Farr, race was linked by inheritance or custom rather than skin color, which is why he argues that "men have the power to modify their race," and as result have the power to change their current actions.[11] Farr's research would inspire Sir Francis Galton and Karl Pearson's work on eugenics, which was later established as a scientific discipline. Galton states: "No more than there is equality between man and man of the same nation is there equality between race and race. This

10 Thomas Malthus, *An Essay on the Principle of Population, As It Affects the Future Improvement of Society with Remarks on the Speculations of Mr. Godwin, Mr. Condorcet, and Other Writers* (London: J. Johnson, 1798), 114.

11 William Farr, "Untitled" (paper presented at Fourth Session of the International Statistical Congress, London, 1860), 4.

differentiation of men in physique and mentality has led to the slow but still imperfect development of occupational castes within all civilized communities."[12]

Vitalism and the Eugenic Descent

Zuberi writes that numerical analyses of phenotypic characteristics are inextricably linked to the field of applied statistics, particularly within histories of eugenics. By promoting the application of scientific evidence to human evolution, eugenicists were able to support claims of racial inequality. This materialized into the application of intelligence tests, involuntary sterilization, and engineered relationships between "desirable" men and women to preserve hereditary lines. Since Black children were thought to lack physical, mental, and temperamental balance, miscegenation, in this sense, was seen as a racial error that could be corrected through the encouragement of racial segregation.

In many ways eugenics originated with Charles Darwin, whose concern with natural processes developed into the notion that humans are part of a larger system of organic life and are subject to the same natural processes of evolution, growth, and decay. Darwin challenged Lamarck's theory in his 1859 text *On the Origin of Species by Means of Natural Selection, or the Preservation of Favoured Races in the Struggle for Life.* The book was not a complete rejection of Lamarck; in it, Darwin combines Lamarck's theory with the work of Malthus. By doing so, Darwin prioritizes the act of survival as the primary motivation for evolutionary development. Darwin applies the concept of natural law to observational patterns of species survival. This later developed into his well-known thesis on natural selection and interspecies competition.

Darwin believed that this characteristic was essential to the development of new subjects. G. R. Searle argues that "the

12 Galton in Zuberi, *Thicker than Blood*, 27.

hypothesis of natural selection was bound to raise in many people's minds the possibility that, as rational beings, men might learn to control their evolution; in place of the blind processes of natural selection."[13] The deliberate effort to "improve" the species was, in this way, a profoundly ambitious project that had to align with the problems being experienced by various industrial and political institutions. Searle notes that in Britain in particular, eugenicists asserted that it was possible to apply the physical sciences, such as biology, to enhance national power and national regeneration—a proposition that directly applied the laws of the physical sciences to the engineering and organization of the human species. Eugenics used statistical method to rank humans on the basis of certain morphological characteristics and social measures. These histories formed ambiguities that underlie the racialization of statistics and, as Zuberi notes, are reminders of the logics of mathematical processes within the social.

Although ideas of race perfection were seen as universal properties, the notion of regeneration relied heavily on the control of hereditary markers that were viewed as undesirable. In other words, as Searle states, eugenicists thought it possible to "breed out" certain grave hereditary ailments in the same way biologists had learned how to breed various fungi and diseases out of wheat. Notable theorists such as David Hume, William Paley, and Jean-Jacques Rousseau adopted Darwin's thesis for further inductive studies of human social behavior. These theories significantly contributed to the establishment of individualist ideologies, as well as laissez-faire political structures.[14]

---

13 G. R. Searle, *Eugenics and Politics in Britain: 1900–1914* (Leyden: Noordhoff International Press, 1976), 4.

14 See Zuberi, *Thicker than Blood*.

Normal Distributions

Johann Gauss's contributions to the field of mathematics are significant when we consider the social impact of the naturalization of normalized standards through prototypical logics. Gauss's development of the normal distribution curve (or Guassian curve) collated random data into a visual representation of their spatial distances from a normalized average, represented as a straight line superimposed on the set of chaotic data. The graph is predicated on a mathematical principle that most data tend to organize around a set of central values, bringing order to what seems like the chaos of unstructured properties. As a technique, the Guassian curve not only subsumes the contingency and independence of living factors into a visual representation of averages, but it also simplifies the logics of eugenics with a universal representation of normality, whereas any individual deviation from the average characteristics of being—i.e., Anglo-European males—could be articulated as their naturalized distance from the prototypical ideal of species. The theory assumes an equivalence between ideal representations of species and statistical averages—challenging notions of data, agency, and the weaponization of mathematics in discursive justifications for the reduction of life as functions of normalized patterns. Importantly, the theory was also applied to phenotypical attributes. While "there is no continuous passage away from the norm," Hacking writes, "if there is, it is to be corrected, the contractor reprimanded, the workman dismissed."[15] From François-Joseph-Victor Broussais's principle of the "normal state" (which represented non-inflamed states of organs) to Auguste Comte's assessments of certain mental states as more normal than others, concepts of normality produce social and scientific understandings of abnormality as attributes to be eradicated, corrected, or manipulated into standards of purity and recurrent return: "The normal ceased to be the ordinary healthy state; it became the purified state to which we should strive, and to which our

15 Hacking, *The Taming of Chance*, 165.

energies are tending. In short, progress and the normal state became inextricably linked."[16] For Comte, however, "progress is nothing but the development of order: it is an analysis of the normal state."[17] As such, Comte's positivism established a social and political framework for the installment of normality as the metaphysical presence perfection, from which all others could be measured and corrected. Comte's alliance with normal distributions informed subsequent disciplines, such as sociology.

For instance, Émile Durkheim sought to improve upon the preservation of good health by institutionalizing normality as a function of observable social behavior. Durkheim would later apply Galton's symbolic language to collective human behavior. In his 1893 text *The Division of Labor in Society,* Durkheim writes: "The only characters transmitted regularly by heredity in a given social group are those whose reuniting sets up the average type."[18] Durkheim was picking up on the empirical scaling of species variation articulated by Galton. For Galton, normalities were something to be improved upon, as ameliorations of the averages of human condition. There was an epistemological character to the understanding of being that constituted a framework of normality as the antithesis of error, devaluing the capacity of deviation to individualize into contingent forms of being and becoming. Measures of deviation allowed individuals to be classified with regard to mean attributes. This is, however, where Galton and Quetelet's theories of species diverge.

Instead of a law of normal distribution and error, Galton emphasized a newly constructed language of human scaling based on theories of species deviation, variability, and survival. However, the juxtaposition of this research with that of his first cousin, Charles Darwin, and Quetelet, resulted in symbolic languages focused on laws of species adaption, specifically biological improvement. Unlike Quetelet, who viewed the normal

16 Hacking, 168.

17 Comte quoted in Hacking, 168.

18 Émile Durkheim, *The Division of Labor in Society*, trans. George Simpson (Glencoe, IL: Free Press of Glencoe, 1933), 324.

distribution of human characteristics as a result of a large number of variable, independent, and indescribable causes, Galton based his assertions on the presumption that species variability was a hereditary characteristic. His formulations led to new mathematical laws of regression and standard deviation still in use today. According to Desrosières, this marked the end of attempts to eliminate species deviation in favor of a greater focus on the statistical means of human attributes, premised on evolutionary survival and symbolic ordering.

Despite the explicitness of Quetelet's theories, the robustness of his methods has been called into question. Doubt was already being cast on Quetelet's claim that a coefficient could be established to express the relationship between small samples and large data sets. Dutch official Baron de Keverberg heavily criticized ideas that a single law of population estimates in the Netherlands, which Quetelet had attempted, could be applied to the entire population. De Keverberg's criticism concerned the heterogeneity of the population and the complexity of factors determining the attributes studied. In short, de Keverberg did not believe generalization could be achieved by using complex systems.

Quetelet's distributions, although rigorous in their creations, did not test data for goodness of fit. In other words, through Quetelet's proofs, linear relationships are established based on a scattered set of binomial data points. This relationship produces a mean, average, or nominal of all of the collected data. Theoretically, this line can be extended into infinity with any future data points predicted to fall within its reach. However, the linear relationship is never an exact match, only an estimation based on a "best fit" of all of the data points. The linearity visualizes certainty onto uncertain distributions.

Mitigations of uncertainty in complex systems were further formalized by Siméon Denis Poisson. Since the institution of jury trials in 1789, lawmakers had become increasingly frustrated with errors in jury decision. These had serious consequences, as an innocent person found guilty could be wrongly sentenced to death and executed. The elimination of error proved difficult as the guilt of the accused is often uncertain, not to mention the fallibility of jury decisions. In settling these concerns, Poisson's experimentations with jury decisions led to his 1837 book *Law of Large Numbers*, in which he theorized a constant law of the same name. Poisson drew on Daniel Bernoulli's "law of large numbers," in which the probability of an event happening in one way or another converges toward a normalizing law, given the number of runs of the experiment increases indefinitely.[19] Poisson, in following Marquis de Condorcet and Laplace before him, applied the technique to generalize jury decisions. In other words, the law of large numbers applies statistical tests to deduce rare events, like jury decisions, into probabilities of certain outcomes. Whereas Quetelet's distributions had not considered variation, Poisson's proofs made it possible to incorporate variability into generalization. His contributions were furthered by his formulation of generalized linear regression models, or what is known as Poisson regression, that allowed for errors in count data models of other than normal distribution.

The debate is founded on the idea that a constant cause creates equivalencies, which then allow events to be aggregated, giving the appearance of contingent evidence functioning through a more general law. Nonetheless, Desrosières maintains that Poisson still considered the subjective by viewing probabilities as degrees of belief or judgments by rational individuals. Karl Pearson, however, rejects the idea of causality in favor of contingency.[20] Adopting perspectives by Quetelet and Galton, Pearson places emphasis on the asymmetries of mathematical representations. The representations provided methods to find fits for skewed (or abnormal) curves, attesting to an underlying homogeneity and constant cause. Although highly contested, Pearson's incorporation of antirealist knowledge informed his interpretation of statistics as a necessary demonstration of the value of correlation. Here he accounts for turns to mathematical law by interpreting symbolism as an available shorthand

---

19 See Kevin B. Korb and Ann E. Nicholson, *Bayesian Artificial Intelligence* (Boca Raton, FL: CRC Press, 2011), 6.

20 Desrosières, *The Politics of Large Numbers*, 105.

when necessary. Desrosières argues that Pearson's departure from his contemporaries was an attempt to capture both realist positions at certain moments as well as degrees of nominalism when discoveries inside of larger systems are needed.[21]

On the other hand, Galton's interest in social ordering extended from physiological characteristics to individual and collective conditions, as a function of heredity. In his book *Regression towards Mediocrity in Hereditary Stature*, Galton circumvents the economic, social, and political forces that were (re)producing poverty in late nineteenth-century London in favor of empirical data. For instance, he compiled social data on poverty into statistical tables (what he called "poverty tables"), which scaled individual economic status according to a series of arbitrary indicators such as standards of living and profession. The correlations were later projected onto a matrix of social aptitude that defined civic worth as functions of genetic character. These precedents would later inspire empirical scales of human intelligence, such as the Binet-Simon scale, Spearman's general intelligence tests for individuals (the "g factor"), and intelligence quotients (IQ).

Objective Frequencies or Transcendent Law?

Zuberi asserts that our view on statistics must be approached from deracialized perspectives to help increase understandings of variations within populations. He writes: "If we deracialize our analysis we could examine the cultural, biological, and social factors that affect population differences without the mystery of race. If race is not biological, then it is not a good proxy for understanding biological processes. If race is, as I have argued, a signifier for the impact of racial stratification, then we may well learn much by developing better measures of social and economic processes."[22]

---

21 Desrosières, 69.
22 Zuberi, *Thicker than Blood*, 142.

Desrosières argues that "the aim of statistical work is to make a priori separate things hold together, thus lending reality and consistency to larger, more complex objects. Purged of the unlimited abundance of the tangible manifestations of individual cases, these objects can then find a place in other constructs, be they cognitive or political."[23] It is plausible that a deracialized approach might alter the relationship between statistics and power. At minimum we might intervene in the hidden relationships between the "two dimensions of knowledge and action" that oscillate between reflections on progress and generation of knowledge for power and control.[24] As Foucault reminds us, "it is not simply the oddity of unusual juxtapositions that we are faced with here. We are all familiar with the disconcerting effect of the proximity of extremes, or quite simply, with the sudden vicinity of things that have no relation to each other; the mere act of enumeration that heaps them all together has a power of enchantment all its own."[25] While statistical methods enrich the field of machine learning, they are fragile to the power of enchantment. Otherwise, as Donna Jones argues, life under these conditions "become[s] nothing more interesting than a specific kind of information in an information age."[26]

23 Desrosières, *The Politics of Large Numbers*, 236.

24 Foucault, *Order of Things*, xix.

25 Foucault, xvii.

26 Donna V. Jones, *The Racial Discourses of Life Philosophy: Négritude, Vitalism, and Modernity* (New York: Columbia University Press, 2012), 2.

ACT III
($\sqrt{-2}$) Toward a Metaphysical Correction of Simple Substances and Ideal Things

CHAPTER 6
The Radical Origination of Divine Black Objects

> Indeed no, the good and merciful God cannot be black.
>
> —Frantz Fanon

If, as Luciana Parisi posits, algorithms generate their own lines of reason toward problem-based solutions, then the study of race is also haunted by the solemn occasion of conventional rule-based procedures. In the traditional sense, algorithms are defined as a precise specification defining the manner or means by which a technical or social problem might be solved. Solution is predicated on the arbitrary designation of a simple or complex question affected by any perceived impairment of a normalized function or process. The problem is a presage of something ill to come, distressing the complex methods governing the relational objectives of the normalized system. We must remember that the system in itself extends beyond the technological and into the organizing metastructures of systemic classification, be they individual or group recognition, behavior, or artifacts that require observation in order to satisfy referenced values. Whether human-made or natural phenomena, a group of independent but interrelated entities comprise a unified whole that works as an arrangement of iterative yet contingent processes. Here, the settlement of performances is carried out in a manner of spatial and temporal circumstances, monitored by an application of possible actions within any given continuum.

Concerning the techno-human, the reconciliation of an algorithmic solution is a settlement of debt, or an assumption of ownership over a set of values that designate truth-based conclusions. But all social relations are marked by a degree of companionship with others. They are composed of a commitment toward a sense of belonging to the prescribed characteristics of the relation. The constitution of the individual and collective, as a recurrent act of companionship, is therefore a vital property of holding together the essential attributes of a communal shaping. This includes the nebulous configurations of recognition, representation, appearance, and other spatiotemporal arrangements that collectively work to compose the substances of a racialized environment.

Nonetheless, as Sylvia Wynter suggests, the abundance of nourishment achieved by abstract social properties is no more than the quality and condition of social hierarchy, built solidly on the formal event of humanism and the dispossession of racial value. This is to say, what is constituted as a "lawlike correlation between our modes of knowledge production" and social praxis is linked parasitically to the fantasies of hierarchal systems, in this case arranged by the substance of racial calculation.[1] Through the formalization of racial category, the constitution of the subjective is descriptive of the various codes and utterances of the racial imaginary. The imaginary, as a statement of colonial fact concerning the superiority of whiteness in individual and collective life, is entangled with a wider circulation of thought that sets the precondition for a necessary correlation between race and a hierarchy of values. This recurrent yet hellish zone of circulation reaches beyond the mystification of racial merit, rendering the individual desirable, valuable, or even useful depending on the parameters of a given hegemonic problem. In this way, the principal values of the racial systems, as an ideal accepted by the colonial order, hold dear the appraisal of racial attributes as a determinant of jurisprudence, whereby the collective perception of racial value, assessed by phenotypical classification, is generalized as an enforceable principle of relational order. Yet, this generalization also concerns a relation between the principles of natural phenomena, the maxims of complex social systems, and the role of basic truth. The basic truths of order, and the laws that impose authority over their procession, produce standards by which the presumption of categorization becomes self-justifying of certain modes of decision-making. But the uncloaking of this fraud is accompanied by a desire to enter into what Stefano Harney and Fred Moten describe as

1 Sylvia Wynter, "The Ceremony Found: Towards the Autopoetic Turn/Overturn, Its Autonomy of Human Agency and Extraterritoriality of (Self-)Cognition," in *Black Knowledges/Black Struggles: Essays in Critical Epistemology*, ed. Jason R. Ambroise and Sabine Broeck (Liverpool: Liverpool University Press, 2015), 203.

a centrifugal force of intimidation.[2] The force, in its unmasking, reveals a paradox in the act of disclosure, which reeks of a marooning desire to validate one's existence as a problem in itself. While to show, or make visible, is thought to catalyze certain alternatives to living through the mobilization of one's (dis)agreement with racial judgment, the birth of a seemingly liberatory antiphon is not the beginning one might expect. It is a prescribed kinship with, or a type of compulsion toward, the arrival of entryism into fictive classification and the more generalized substances of racial order.

Ontic Vertigo

It was Du Bois who asked how it feels to be a problem. The "problem" Du Bois alerts us to is twofold.[3] There is the question of Black presence as a state of inconvenience within a predominantly white system, which finds serious difficulties reconciling the proximity of difference (as it currently exists) with the ideological statement of desired racial purity. Then there is the problem of Black being as an internal condition always already negated by the (in)visible impressions of everyday racism. The state of affairs that brings this problem of being into question produces, in first instance, a psychic alienation that emerges and reemerges with each utterance of assigned racial value. For Fanon, this psychodrama is at the heart of the ontological misrecognition of Black existence and the maturation of individual and collective relation—a problem he encloses in the notion of sociogeny. Without immediate recognition of the mnemonic residue of racial consciousness, the scenes from which the act of psychic trauma are provoked reproduce an arbitrary yet enigmatic disturbance of systemic dysfunction. Here, race maneuvers

---

2 Harney and Moten, *Undercommons*, 95.

3 Du Bois, *The Souls of Black Folk*, 3.

out of and in between a collective unconscious, as Fanon pulls from Jung, enacted in the everyday praxis of being human.[4]

Autopoetics and Our Release from Epistemic Madness

But not all is lost. Given Fanon's sociogenetic schematic and the reemergence of the ontological question of the post-human, Black studies has returned to the colonial episteme, the correlation between data, the history of racial coding, and invention of the Other. Fanon had already situated the problem of race as a problem of exclusion, as well as the arresting of Black psychic potential. For Fanon, any ontological consideration of psychic genesis, human development, or social praxis is epistemologically deficient if it does not recognize racism and colonialism as the cause of the disproportionate relation between social development and the Black psyche. Perhaps to recognize this desire to convene with the event of race, as well as commune with the experience of self, is to toe the line of racial substance as always already negligent and incapable of providing the nourishment we so desire. It is instead to entrap disclosure into a realm of incompatibility with self-perception in order to catalyze alternative perceptions of one's relation to self through and in between the undifferentiated labor of the racial-epistemic alliance.

Wynter has shown that dreams of disclosure provide spaces for the malnutrition of colonial and imperial violence. The dream further demystifies belief in the rewards of representative organization, while shifting the prophesy of racial order into that of the fantastic. As with any fanciful design, the wondrous howling of colonial utterances is a faint noise within the subjective sense of self, whereby the individual and collective vibrations of speech are transmitted by an elastic medium without concern as to the appearance of comprehension or whether or not it can be distinguished by a particular speaker. It is speech in and of itself, for and by itself, as already in relation to the larger conditions of

4 Fanon, *Black Skin, White Masks*, 122.

transindividual responsibility. Wynter draws on the biological notion of autopoiesis as a way to investigate the legitimating structures of racial hierarchy and what she describes as an "indispensable condition of the formation and stable replication of each respective societal order."[5] Francisco Varela and Humberto Maturana developed the concept of autopoiesis to explain the phenomenon of living organisms and their cognitive capacities.[6] An autopoietic system is an enclosed and autonomous system that distinguishes living and nonliving systems. It describes living organisms as self-producing and the nature of perception and intelligence as subject dependent. It presents a process of production that generates its components in recursive recreation of the self. An autopoietic system is realized in a particular structure and is independent of its environment.[7] A key component of autopoiesis is the relation Varela and Maturana establish between closed recurrent systems and cognition. In general, cognition refers to the assimilation and use of knowledge domains, and as such is limited to beings with complex nervous systems. Both cognition and perception are linked in the operation of the nervous system, which is realized through the autopoiesis of the organism. The survival of the organism, as an autopoietic system, depends on a cycle of recurrent interactions that produce knowledges of the relation, including the interaction between discrete entities within a closed system. In other words, the organization of cognitive systems define domains of relational knowledge and enable action.

Although Wynter grounds her search for being in an autopoetic struggle for recognition, she suggests that any fictive construction of substance is also a self-programming schema

5 Sylvia Wynter, "The Ceremony Found," 203.

6 Francisco J. Varela and Humberto R. Maturana, *Autopoiesis and Cognition: The Realization of the Living* (Dordrecht: D. Reidel, 1980).

7 For a detailed description of Maturana and Varela's concept of autopoiesis, see John Mingers, "The Cognitive Theories of Maturana and Varela," *Systems Practice* 4, no. 4 (August 1991): 319–38.

that enacts a symbolic set of behavioral instructions. As Wynter argues, these instructions are not purely operational, but are psycho-affective in that they form a cognitive closure around fictive typologies. The cognitive closure expresses the boundary of the *we*. In other words, it defines the referent of *we* as that which is not the Other—the "*they* and *not-us*."[8] As such, it magnifies the specificities of type, including the symbolic operation of life, death, belonging, and their preservations. In reference to Louis Althusser's *Ideology*, Wynter contends that what we know as humans and how we experience our environment is framed by sociogenetic code. This means that our knowledge of "reality" of the autopoietic living system, as well as coordinated behaviors, enacts a "perceptual categorization system" that is incompatible with the way reality is viewed outside of genre-specific viewpoints.[9] Wynter argues that bodies within any autonomously functioning autopoietic living system typically have no direct "cognizing" access outside of the terms of the system.[10] In other words, by creating a higher level of the divine, subjects are rendered ignorant of high-level systems of knowledge. Here, the subject can only perceive what is enacted in relation to other secular beings. For Fanon, Wynter writes, this empirically derived self plays a central role in the criteria of being, which gestures toward the colonized psyche and experiences the self as if it were actually genetically inferior. She writes: "For it is on the template of these marks/criteria and the governing codes of symbolic life and death [...] which they express, that all individuals can alone be socialized as the condition of their realization not only as culture-specific subjects, but also as ones able to experience themselves as symbolically conspecific with the other members of the 'we' with whom they are narratively/linguistically bonded as they are biologically programmed to be."[11]

8 Wynter, "Human Being as Noun?," 43.
9 Wynter, 45–46.
10 Wynter, 72.
11 Wynter, 45-46.

Here, Wynter points to Fanon's dispute with liberal humanism and the premise of bioepistemic human development. It is by means of this strategy that the process of socialization and regulation of individual beings and collectives become central concerns for the ordering of human and culture-specific modes of reality. The discretization of the Black body into accumulations of calculable elements is intrinsic to the formation of an empirical reality as a political project of visibility. Under these conditions we might also think of negation as a process of white prototypicality, or a standard that epitomizes whiteness as the ideal form of species, and therefore the prototype for Black being. Lewis Gordon turns to Wynter's interpretation of Fanon to illustrate the pathological strain Blacks experience under the epistemic relation and the dialectic of recognition, which fully materializes once being is substituted for visibility and physiology.[12] Gordon posits that Blacks become narcissistic when confronted by their own self-image in relation to the standard model of white representation, which also reinforces Black beings' desire to be and behave as white people would. As Gordon writes: "[Blacks] thus live with the knowledge that the world is larger than the white one, and they know that the ascription of being is not granted to that wider world—that world of, as it were, dark matter—but they also know that they live in that world, it is their lived experience. Whiteness exemplifies a kind of blindness. It is a patronizing view of blackness as a limit, a limit of being, a point of lack."[13]

By applying an autopoietic schema to the colonial imaginary, Wynter attempts to grasp the layered patterns of global knowledge systems that function as perceptual systems of categorization. The instrumentalization of this fictive concept is enacted by the colonial episteme, which, as Wynter posits, is a

12 Lewis R. Gordon, "Is the Human a Teleological Suspension of Man? Phenomenological Exploration of Sylvia Wynter's Fanonian and Biodicean Reflections," in *After Man, Towards the Human: Critical Essays on Sylvia Wynter*, ed. Anthony Bogues (Kingston: Ian Randle Publishers, 2006), 237.

13 Gordon, 240–41.

familiar concept in the history of European social systems. This prolonged eugenic descent not only legitimizes the systemic circulation of colonial utterance, but it also substantiates the ontological and hegemonic status of global class structures. For instance, the European feudal systems developed in the eighth century, along with the feudatory relationship between lord and sovereignty, vassal and allegiance. The owning or owing of allegiance, of pertaining to being as a subject of sovereignty, was not solely dependent on physical service but also on the entitlement of benefit and/or loss as an act of availability for use by an external authority, either literally or figuratively, within an established arrangement of categorical order. It is widely known that a condition of this specified arrangement is marked as suitable to the equilibrium of a system, as well as the continued maintenance of the categorical order, by establishing a logical comprehension of a decree that binds the relationship into a regular arrangement of separate, functional qualities. Here, the comprehension of one's place within society becomes synonymous with the status of being ordained to the office of sacrament and social tradition. As Wynter describes, the formal ceremony of one's conference into this divine order, as accepted by a specific grace on those who receive it, grants a qualification, title, right, or possession by or of another in and of external ordinance.

Divine texts were the main source of knowledge in feudal systems. By the fifteenth century, however, divine knowledge had been largely disputed by humanists, who believed that one's full potential could be reached through education in classical literature and the promotion of civic virtue. As values shifted away from the divine and toward more secular interests, so did ontological assertions about human dignity and human capacity for fulfilment through reason and scientific method. The humanist belief in the continuous emergence of evolution rejected religion as pertaining to human abilities to learn knowledge and understand the facts and significance of their behaviors. The tolerance for the retention and reproduction of this knowledge amounted to a strong faith in verifiable acts of scientific and mathematical phenomena, forming the means whereby control over human destiny was tantamount to the amount of information that could be acquired and scientifically validated. The resultant atmosphere redistributed the power of a single divine provenance, substituting it with a practical subject that was organized according to the principles of cognitive reason and technological development. The process of individual improvement—as a complex expansion, enlarging, and refining of human capability—effectively merged the biological unfolding of events (and with that the gradual changing of the world) with the maturation of a state of purpose (delimited by an extensive search for information that could sufficiently explain the elaborate composition of humans and the environment alike). For Wynter, the product of this extensive network and circulation of data, extracted from an active site of human-centered knowledge production, forecloses on the remarkable phenomenon of human variability by instead bringing forth a state of knowing through intuition or reasoning rather than lived experience, apropos preexisting racial categories. Although technological validation had usurped divine holdings, the categories of human racial difference remained intact, thereby translating the biological from theological to an a priori organization of scientific principles.

Both Wynter and Fanon see the attribute of race as an epistemic set of conditions that absolve science of racial assumption. Here, science declares its innocence while simultaneously promoting notions of racial difference. The enactment of the episteme, or what Foucault describes as the coexistence of a set of relations that form the conditions of possibilities for categorical hierarchy, inform our understanding of racial attributes as parts of a whole yet single substance.[14] The episteme is a means by which the Other is not only brought into being, but brought into coherence by science as difference in itself. As Foucault has shown, the episteme operates under discrete forms of mundane practices and solutions. Foucault initially restricts the episteme to the distribution of scientific knowledge as a mode of power,

14 Foucault, *The Archaeology of Knowledge*, 211.

but expands the concept in later writings to account for other knowledges produced outside of science that remain invisible, concealed, or "epistemological[ly] unconscious."[15] Here, the individual of difference embodies the normalizing power of falsification, formed and conforming to the applied study of experimentation and observation rehearsed through rigorous discipline of problem-based domains.

Wynter argues that this ultimate mode of Western being—a mode of non-Otherness based on race—generates additional forms of subtypes that are logically measured, ranked, and bifurcated based on any ethnophenotypic distance from normalized models of Western culture. For instance, the construction of the lower social classes as measured against the middle classes, non-heterosexuals against heterosexual standards of behavior, women's bodies in comparison to men's, and so on. These organizing principles are instrumentalized through "lawlike" effects that produce what Fanon observed as an autophobic organization of behavior in his Black patients.[16] Behaviors are regulated collectively on the levels of affect, perception, action, and cognition. Nonetheless, through the origin of teleological classification, Wynter searches for a possible opening to historical concepts of being. She posits that any attempt at situating being must emerge from outside of intra-Western terms, which have already solidified ideas of origin and prototypical distinction. She describes this revised narrative as a planetarily extended story that exceeds the limits of Renaissance humanism, as well as humanism's resurgence as an institutionalized and biologically absolute expression of neoliberal fundamentalism. Her ultimate aim is to illuminate the distinction between the colonized being and the discursively produced depiction of ethnophenotypic people as not-human or nonbeings. The theoretical distinction seeks to separate the *being* of *being human* from the invented concept of *being human* established under the terms of the fictive Man.

---

15 Keith Albert Sandiford, *Theorizing a Colonial Caribbean-Atlantic Imaginary: Sugar and Obeah* (New York: Routledge, 2011), 29.

16 Wynter, "The Ceremony Found," 31.

Wynter states: "The central mechanism at work here [...] was and is that of *representation*. Its role in the process of socialization, and therefore, in the regulation both at the individual and at the collective levels of the ensemble of behaviors—affective, actional, and perceptual-cognitive—is central. For it is by means of the strategies of representation alone that each human order and its culture-specific mode of empirical reality can be brought into being as such a 'form of life' and third level of human, and therefore languaging existence."[17]

While Foucault is aligned with Fanon's articulation of the problematic as one of form, Wynter extends the critique to consider the ways in which substance is anterior to the categorization of race. This race substance, as she describes, is a circular and self-justifying process enacted by categorization and the production of scientific criteria. According to Wynter, bioepistemic proportion adjusts the spaces of knowledge to account for the declaration of scientific truth and, at the same time, works to absolve whiteness from its estrangement with the hostilities of racial categorization. With the notion of the bioepisteme, Wynter seeks to confront the historical externalization of racial violence—namely the recurrent draft of racial hierarchy. Through a durable yet consistent narrative, or "hearsay," as she describes it, the deductive principles of ethonophenotypic necessity are legitimized in their hegemonic social status. Wynter classifies the bioepistemic under four schemata: the biocosmogonic, the sociohistorical, the political, and the salvific. These bioepistemes are articulated in the coding, or "prescriptive sociogenically" encoding, of slave and colonized bodies and in the regulation of their movement.[18] Here, she returns to the question of origin by drawing parallels between fictive narratives of human history and the emergence of the humanities as a research discipline during the Renaissance. Its most significant byproduct was the

17 Sylvia Wynter, "1492: A New World View," in *Race, Discourse, and the Origin of the Americas: A New World View*, ed. Vera Lawrence Hyatt and Rex Nettleford (Washington, DC: Smithsonian Institution Press, 1995), 45.

18 Wynter, 69.

invention and valorization of the secular concept of Man. The humanities then emerged as a damaging orthodoxy in the nineteenth and twentieth centuries into what Wynter describes as "the ossified *Scholastic* orthodoxy of medieval Latin-Christian Europe's order-instituting/order-legitimating theologically absolute answer" to the question of origin.[19] As Wynter and Fanon posit, racial substance has become a recurrent and consistent process. It results from an irregular series of affect, lodged in a "certain uncertainty" within the universal concept of human.[20] As such, the diagnosis of Black alienation is, as Fanon argues, a sociodiagnostic phenomenon that can neither be reduced to biochemical processes nor "escape human influences."[21] The coherence of racialized attributes, or the fictive substance of race, links the dynamic instrumentalization of coherence found in the bioepistemic to the "discursive negation of co-humaneness."[22]

Who Needs God (or Science) on Their Side?

This abandonment and delegitimization of Black existence subtend the divine angle of perception, drawing a boundary between the essential qualities of being human and the full-scale workings of a prototypical model of humanity built for the engineering, testing, and release of white superiority. It is within the construction of this divine grant that Fanon induces a type of ontic vertigo in philosophical doctrines on the study of human experience. Fanon is insistent that the application of theory for observable predictions of human existence, and their considerations of objective reality, are blind to the remarkable way that science is always already capable of determining the limits of humanity. Fanon's call for an "unrepresentative (unproblematic)" account of humanity highlights the ontological misdirection of

19 Wynter, 5.
20 Fanon, 90.
21 Fanon, 4.
22 Wynter, "Human Being as Noun?," 4.

universality.[23] This is why Fanon asserts that the Black man has "no ontological resistance in the eyes of the White man," even if they occupy a certain space of equivalency.[24] The problem thereby disengages the syllogism of white superiority. Fanon's rearrangement of the ontological perspective is an attempt at extricating the terms of the humanist proposition from the copula of race, whereby the premises of race must be analyzed to address this syllogism instead of expressing judgment for the sake of classification by name. Still, Fanon is careful to suggest that the complexities of race relations are too often supplanted by discursive concepts of racial transcendence and racial consciousness—ones that refuse to decenter the white subject as the protagonist in a shared transcendental ego. Ontology must articulate a new mediated consciousness that auto-designates a reality that synthesizes the actualization of the self with the production of the social.

If we build upon Wynter's diagram of human classification, where the human species is classified as either rational or irrational, then the field of relation is also open to the rejection of "ideological unveiling" of a racial dominance that overrepresents the genesis of being as always already delineated by European male perception.[25] Wynter posits that in terms of racialized life, autopoietic, instituted living frees human knowledge of its physical and biological realities and substitutes these realities for order-stabilizing and order-legitimating codes. As such, levels of human knowledge of the physical and biological are emancipated by epistemological processes that circulate in the autopoietic field of relation. Furthermore, an opportunity unfolds to challenge both the symbolic ordering of phenotype, as well as the Hegelian dialectic of recurrent associations.

23 Wynter, 4.

24 Fanon, *Black Skin, White Masks*, 83.

25 Denise Ferreira da Silva, "1 (life) ÷ 0 (blackness) = ∞ − ∞ or ∞ / ∞: On Matter Beyond the Equation of Value," *e-flux journal*, no. 79 (February, 2017), https://www.e-flux.com/journal/79/94686/1-life-0-blackness-or-on-matter-beyond-the-equation-of-value.

The provocation prompts consideration of the contemporary organization of the world. If, within the schematic of sociogeny, the radicalized body is disassociated by the consistency of race substance; if the Black body is comprised of a "thousand details, anecdotes, and stories";[26] if this body is organized under the terms of "History" and "Truth"; and if this "Truth" is instrumentalized as a justificatory act, then what remains of the racialized body in contemporary big data and epistemic processes? This question hints toward contemporary epistemic operations that, once enacted, can subsume the phenomena of existence into consistent organizations of discrete data—that is, operations that are designed to accumulate, fragment, organize, and classify phenomena under illusive yet universal authorities of scientific knowledge. We must question contemporary operations that materialize as "new ways of seeing" and, in particular, increase the capacity for pejorative judgment. The classification of the object relation, as well as difference therein, is a longstanding issues of object compatibility within the autopoietic network of interaction—not only the circulation of seemingly incompatible objects, as in the racial distinction Wynter describes, but also the ontological maintenance of incompatibility as difference in itself.

Given this, we might redefine the epistemic terms of compatibility within a prolonged eugenic descent that owes allegiance to an enduring history of social quantification. The latter being the collection of artifacts from which conclusions may be drawn, be they organic involving human biological process, or having the characteristics of nonliving components. These data, derived from measurement, assign fact to systematic investigations—or, more so, to what is capable of being factual or non-factual within a cluster of observed phenomena. Being, as reflecting the genuine character of fact as it coincides with reality, amounts to the material investigation of measurement as having substance. This idea gains considerable capital in the conveyance of the phenotypical as a uniform property of

26 Fanon, *Black Skin, White Masks*, 84.

being-in-relation to oneself and one's environment. The choicest and most essential of these properties are not to be taken lightly, as they conflate the occurrence of illusion with verified statements of ethnophynotypical existence. In this way, data is an essential component in reflections of race, not to mention the ceremonial fixture of white essentiality.

CHAPTER 7
A Correction of Metaphysics and the Concept of Black Substance

> I do not wish to be loved, I adopt a defensive position. And if the love-object insists, I wish to say plainly, "I do not wish to be loved." Devaluation of self? Indeed yes.
>
> For the object, naturally, need not be there, it is enough that somewhere it exist: It is a possibility.
>
> —Frantz Fanon

On a journey from Paris to Hanover in 1676, having spoken with his contemporary Baruch Spinoza "several times and for very long," Leibniz elaborated on his treatise for the perfect being.[1] In his discussion with Spinoza, Leibniz attempted to settle ongoing debates on the ontological contradiction of existence. In his 1702 "Letters to Varignon, with a Note on the 'Justification of the Infinitesimal Calculus by that of Ordinary Algebra,'" Leibniz argues that to think through existence is to think through the problem of incompossibility. That is, to understand the relation between incompatible aggregates not as their incompatibility, but the magnitude of relation between them as a function of the predestined harmony. Leibniz uses the kinetics of physical matter as an example. For instance, a grain of sand cannot pass through and divide glass. Equally, sand has incompossible relation with the terrestrial globe since it cannot pass through or divide it either; similarly, the terrestrial globe with the outside universe. The grain of sand in this example is a simple substance, which in Leibniz's view extends in aggregate to make up glass as well as the terrestrial planet. In the communique, he argues that any single substance, despite its various aggregates, can be conceived as a whole in and of itself. Leibniz eloquently describes these aggregates as simple "perfections" in a coordinated dance with other part-as-perfections, whereby the characteristics of each aggregate, as a totality, circulates within the single substance

1 Leibniz, "Two Notations for Discussion with Spinoza, 1676," *Gottfried Wilhelm Leibniz: Philosophical Papers and Letters*, ed. and trans. Leroy E. Loemker (Dordrecht: Kluwer Academic Publishers, 1989), 167.

of the divine, "which expresses whatever it expresses without any limits."[2] In other words, these aggregated perfections, although discrete, contain an infinite number of formal rules and calculated choreographies. Despite this harmonious dance of the divine, these perfections maintain their differences in attributes, which for Leibniz opens up an infinite network of contingency set forth by the conditions of the single substance. Since these contingencies are at their simplest determinate, Leibniz takes a shortcut to describe any incongruent relation between them as a compatibility only understood on the level of the single substance. As extensions of the single substance, they are compatible despite the apparent differences they contain. In this way, any patterns or behaviors resulting from glass or the terrestrial planet are extensions of the behavioral patterns of the grain sand. At the same time, the maneuvers are relational and self-organizing—not fixed.

To a large extent, the emergence of the logic of relations is divided into two periods of thought: before and after the mid-nineteenth century. But where do we begin with relation, particularly given that relation differs depending on how it is understood? The philosophical basis of the ancient Greek tradition lies in attempts to locate external relations ontologically. In other words, the project of philosophy was to comprehend the ways things (people, numbers, communities, objects, and so on) are described—in particular, the ways two or more things are related. Of concern were how these things, and the relations between them, might account for difference. Logics of relation exercised enormous influence from the Middle Ages until the late fourteenth century. It was in medieval philosophy that relations became a central philosophical issue, and terms for "relations" (*respectus*, *habitudo*, *proportio*, *relatio*) concretized into a philosophical understanding. Although largely theologically motivated, medieval philosophers developed their own conceptions of relational situations under these terms, specifically in

2 Leibniz, "Letter to Varignon," in Loemker, *Gottfried Wilhelm Leibniz*, 543.

the denial that relational situations can be molded into categories. If Aristotle attempted to define relation from the object onward, or from the accident of and between objects into characterization, then the medievalists attempted the inverse: to start from signification to work toward an explanation of the relation therein. By the fourteenth century, however, the denial of the existence of any discernible distinction between a substance and its accidents became commonplace. This departure left open views on relations as being ontologically grounded substances themselves, which has had a profound influence on contemporary modes of technological understanding. Moreover, Scholasticism defined relation as a minimal thing, or the least of all beings, something tediously small, thus claiming, ontologically, little status compared to substance.

To gesture toward the relation of and between things is to understand relation primarily from the point of view of the object; or, in many cases, the object as an objective determination either inherent in itself or attached to itself with respect to another. This movement is a direct extension of Plato's and Aristotle's thinking. Aristotle's *The Metaphysics* elaborated on the specificity of the categorization of relations as "accidents," or entities dependent on substances. In Aristotlean logic, substances are "primary," meaning that there is a static basis that comprises reality, and exist separately from human perceptions, thoughts, and conceptions of human subjects.[3] Accidents here are to be understood as characteristics (or attributes) that may or may not belong to a (human) subject without affecting the essence of the subject. The primary substance then is distinct from (human) subjects who experience or perceive such an entity. In this way, the primary substance can be thought of as an entity unto itself. Thus, according to Michael Halewood, "the task of science is to provide the correct attributes to substance, and the task of epistemology is to account for how knowledge of the 'correctness'

3 Michael Halewood, *A. N. Whitehead and Social Theory: Tracing a Culture of Thought* (London: Anthem Press, 2011), 10.

or 'incorrectness' of such attributes is possible."[4] In his short treatise *Categories*, Aristotle made early attempts to categorize relations on the basis of logic by suggesting a methodology to explain the "truth" of relational events or situations (also called "relational situations"). It is then no wonder that the distinction between things thought to be what they really are (primary substances) and their qualities or characteristics (heat, height, weight, or criminality, for example) would ground themselves in both scientific and philosophical priority.[5] Although relations were thought to be one of the lower types of categories (*summa genera*), it is worth remembering that Aristotle lists them as one of the ten highest kinds.[6] And however fruitful efforts were to combine subjects and predicates together to form simple propositions—what would later be defined as categories or categorical propositions—these relations or accidents amount to properties, inhering in things, of being-toward-another.

How these paradigms transfer into the logical domains and linger in modern thought is profound. According to Halewood, the influence of Aristotlean logic in contemporary conceptions of science and technology studies has had an enduring influence on notions of nature as "inert, passive, and dead."[7] Citing A. N. Whitehead's critique, Halewood asserts that our modern scientific point of view arises from the "slow influence" of Aristotle's logic, which declares that rational and logical thought is based on, as well as expressed through, its simplest form. It has also fueled a largely unquestioned acceptance that underneath any sense of awareness is a substratum, or sense of concreteness, on which scientific knowledge is or can be based. Halewood argues that this "leads to the conception that it is the attributes

4 Halewood, 13.
5 Halewood, 13.
6 John Cook Wilson, "Categories in Aristotle and in Kant," in *Aristotle: A Collection of Critical Essays*, ed. J. M. Moravcsik (London: Palgrave Macmillan, 1967), 74.
7 Halewood, *A. N. Whitehead and Social Theory*, 10.

of matter rather than matter itself which are present to, and the basis for, human perception, knowledge and consciousness."[8]

Halewood writes further that the problems associated with the above position are threefold: First, the universe is conceived as divided into two distinct realms, that of substance and that of the qualities of substance, which are separate and abstract from the substance itself. Second, the interconnection between individual entities of materiality cannot be either accounted for or allowed. Here, the relation between substantial things is disallowed, as each is conceived of as complete in itself. (Halewood argues that the description of facticity limits the description of entities as solitary, unrelated, and situated as individual points in space and time, therefore making it impossible for information to pass between entities—or to account for any mode of dynamism). Third, the bifurcation of the primary substance from those subjects that experience the substance creates a gap in knowledge, where the primary substance becomes that which knows (the knower), and the subject becomes that which is known. In other words, "a gulf is thus created between the world-as-it-is and information about the world-as-it-is."[9]

Oresme, Object Positions, and the Linear Proposition of Race

The problem of the discreet object (or, here, the racial object) has been one of measurement and calculation since the twelfth century, when Nicole Oresme introduced the straight line as a symbolic representation of the relation between time and spatial distance. Prior to that, symbolic functions were seen as providing more general insight into uniform motion between discreet points—as opposed to the traversal of intervals. Uniform motion was critically studied by Aristotle, Autolycus, and Archimedes. Aristotle in particular was concerned with the relation between proportional distances and proportional time.

8 Halewood, 11.
9 Halewood, 13.

Oresme, on the other hand—having studied Aristotle's *Ethics*, *Politics*, *On the Heavens,* and *Economics*, among others—formed a critique of Aristotle's ideas of motion, which Oresme would later use as the basis for his developments in mathematics, philosophy, and science.

Oresme is one of the most eminent Scholastic philosophers, made famous for his critique of several Aristotelian principles. Oresme's rearticulation of Aristotle's concepts of motion were greatly influential in the developing field of mathematics. In his commentary on motion, Oresme diverges from Aristotelian physics by arguing instead that the physical place of any body is defined by the space the body fills or occupies. For Oresme, space is not a substance or an Aristotelian accident, meaning it cannot be wholly signified through language. Space is better understood metaphysically; not as a noun or pronoun, but instead an adverb like "here" and "there."

Oresme's commentary on Aristotle's *Physics* is of particular importance with regard to his views on time and the ontological status of accidents. In Aristotelian physics, the passing of time is proof of the existence of motion. Time here is defined as the measure of motion between points before or after a specified origin. Time and motion, in this sense, are necessary and interdependent. Oresme, on the other hand, defines time as a continuous series of successive durations of things. This continuity of discrete durations is independent of motion (or *fluxus*). Oresme's duration, *duratio rerum tota simul*, is derived from the eternal relation with the divine, and is therefore durable and consistent. As such, Oresme's *fluxus* is also a universal expression of the divine (and thus applicable to all concepts of change) and would foreshadow what we know today as classical physics. Oresme's view of space, however, is closer to that of Isaac Newton's. For Newton, place or position is merely the approximation of substance. Oresme nonetheless extends Newton's theory to describe the origin of space beyond the world we experience, or extra-cosmic space. For Oresme, there exists an infinite void space in the extra cosmic, which he describes as God itself. This became a defining feature of Oresme's philosophy of nature and development of mathematics.

With the divine in place, Oresme expands his concepts to the positionality of discrete units in space. He uses the rectangular latitude and longitude coordinates and the resultant geometric figures to distinguish between uniform and nonuniform distributions of values. Here, we can think of the change in velocity in relation to time, or the distribution of the magnitude of quantities in relation to things. From this starting point, Oresme devises his concept of uniform acceleration, which he represents with two-dimensional right triangles and three-dimensional figures. This new measurable function of velocity and spatial distance helped form the basis of calculus, which made it possible to determine distances traveled, rates of speed, or even transactions per second using symbolic representation.

Calculus also provides a point of reference from which discreet values can be depicted and kinetically preempted. Oresme represents this function with what we know today as the straight line. Oresme's graphical representations of motion further informed developments in kinematics. Kinematics increased in popularity through Oresme's critiques of Aristotle, and further with René Descartes's bridge between algebra and geometry. The concepts, however, were limited by their dependence on linearity and continuity, even as they brought rationalism into greater prominence in mathematics. Leibniz built upon the centrism of linearity. He was the first to see that linear coefficients were not confined to linear equations, and instead could be manipulated and arranged into an array, or matrix. They were highly influential for Galileo, as well as for Descartes's development of analytic geometry. It was Descartes who brought mathematics closer to the notion of continuity between discrete objects.

For Descartes, however, continuity traverses the symbolic into notions of the "real" world. Here, the continuity of real objects in motion is a uniform phenomenon that can be represented visually as a geometric object in space and time. He argues that symbolic functions can not only represent kinetic phenomena, but the representations themselves can be brought into visibility and abstraction as indications of the real. The postulate is as such: if a symbolic function can be represented as a visual representation, then a visual representation can (reciprocally) be represented by a symbolic function. Therefore, the visual representation, as the common link between the symbolic and real, can be dispensed altogether, leaving a pure continuity between perception of the real and symbolic function. Put another way, what is perceived as real can also be perceived as the calculable properties of the world around us, which are made visible by the laws of motion. What remains, then, are not discreet objects that make up events or even the events themselves, but the symbolic functions of the events, which are comprised of smaller, infinitesimally discreet behaviors. Experts in signal processing and machine learning, like Mark Davenport, state that in order to combat these complex problems, the mathematics of algorithms need to be understood in order to identify potential shortfalls, which he defines as the "significant gaps between the abstract problem formulation and what you encounter in a real-world application," which is also beneficial for research into advanced perception and control for autonomous cars.[10]

Differential Segments and Preestablished Harmonies

Leibniz's deviation from Oresme's "ordinary algebra," which relies on linear causality, finds relevance in contemporary kinetic research, including cognitive science and machine learning devices like general adversarial networks or less complex types of neural networks. Along with Boolean algebra and symbolic logic, Leibniz developed differential calculus, disputably independent of Newton, which is a more dynamic formulation of calculus. Differential functions are important in the history and philosophy of mathematics, and they are also crucial to our understanding of contemporary machine learning. For instance, the sigmoid function, which is used in a variety of applications including artificial neural networks, is differential—it makes use

10 Mark Davenport, quoted in T. J. Becker, "The Mind of the New Machines," *Horizons*, March 8, 2018, http://www.rh.gatech.edu/features/minds-new-machines.

of differential calculus for greater operational efficiency. One of Liebniz's most significant motivations behind differential calculus is the ability to create linear estimates of otherwise dynamic points in space and time. Leibniz's ontological concern is in the dynamic relation between the part and the whole in domains of space and time. In *The Monadology,* Leibniz rejects Descartes's materialism, or the view that everything exists in physical form. Descartes's principle favors matter over mind. For Descartes, the mind is merely an instrument that the individual uses to ascertain truths in their relation to the world. For Leibniz, however, the mind and the physical form are two distinct, nondualistic domains. Leibniz argues that the mind—which includes perception—cannot be explained mechanically, and thus cannot be purely biological processes. Instead, the mind and any perception of the world must be one single domain or conscious being of "I." He suggests that the mind and perception are extensions of a single "perfect" substance.[11] In this context, any relation between perception and the world—which he argues is comprised of distinct parts—is also an extension of a single substance.

An important component of Leibniz's claim is his articulation of the relation between distinct parts in a whole domain. Leibniz accounts for the relation between distinct parts (as well as their characteristic differences) by arguing that while appearing to be self-organizing, they are, on the contrary, interacting in a preestablished harmony, which is set forth by the simple substance of the divine. While Leibniz is explicit in his mind-body theory, he is also not shy in describing the issue of infinite difference. As a perfect harmony, incompossibility is thus an infinite coordination of divine provenance that Leibniz defines as the "best possible" relation between bodies. Mathematically, incompossibility as harmony would mean that any distinct variable X and Y, although contingent in themselves, are also situated in an infinite number of preestablished relations that only need to be

11 Leibniz, "Critical Thoughts on the General Part of the Principles of Descartes, 1692, On Part I," in Loemker, *Gottfried Wilhelm Leibniz*, 387.

discovered by science. As such, the perception of nonexistence does not preclude the fact of its existence; and therefore, any formal scientific reasoning of its existence can be taken as a real condition. Leibniz explains in terms of the straight line: "Even if someone refuses to admit infinite and infinitesimal lines in a rigorous metaphysical sense and as real things, he can still use them with confidence as ideal concepts which shorten reasoning, similar to what we call imaginary roots in the ordinary algebra."[12]

While Leibniz's monadic theorem attempts to explain variability as an organized system of relation, there is little justification for the assumption of a singular substance. In doing so, Leibniz stages the single substance as the precondition for imagination, while at the same time arguing that the imagination rests outside of empirical reason. He solves this conundrum by returning to the notion that all matter is a necessary extension of the real. He writes: "Even though these are called imaginary, they continue to be useful and even necessary in expressing real magnitudes analytically."[13] What remains is the pure first value, which is the relation of difference as change itself—a concept Leibniz articulates mathematically as the rate of change. According to Leibniz, any variability is scaled throughout the universe on orders of magnitude. Using his example, we can imagine the variability of a particular entity as a grain of sand, within which it exists all perceived variability. When compared to, say, a single beach or the span of an ocean floor, this variability can be imagined as inconsequential as it vanishes into the larger meta-kinetic system.

Consequently, Leibniz substitutes the imaginary and any conception of the real for a differential calculus with self-defined magnitudes of relational importance. If an imaginary can be conceived, its empirical value is significant inasmuch as it can be expressed as instrumental reason. This applies to the simple grain of sand, as well as any physical structure or assemblages in between. Within Leibniz's legacy, if data can be conceived of

---

12 Leibniz, "Letter to Varignon," 543.

13 Leibniz, 543.

as the simplest indivisible body within a single substance, then any imagined associative aggregates or incompossibility among them can be viewed as extensions of preestablished relations that stand with universal patterns. According to Leibniz, our dynamic individuations are "well founded fictions" that can, in theory, be replaced by imagined quantities of change.[14] Leibniz not only formalized the correlation between biological phenomena and statistical patterning, he also created differential calculus.

Today, advances in machine learning techniques have helped accelerate the development of other techniques that benefit from differentials' agility in data-rich environments, as they did during the golden age of statistics in the seventeenth century. Whereas most statistical and early machine learning techniques are useful in linear environments, they are less adept at classifying nonlinear data structures. A data structure is a particular way of organizing data so that it can be used effectively. A data structure is said to be linear if the elements form a sequence, such as an arrays, lists, and queues. In contrast, nonlinear data structures do not form a sequence, which makes the data more useful for binary trees. Binary trees, for example, are useful for forming data hierarchies, often in parent-child relations. The principal aim of data structures is to reduce the space and time complexities of different tasks—space complexity being the total space taken by an algorithm with respect to its input size, and time complexity being the estimated time it takes for one iteration of an algorithm.

If we can conceive of Leibniz's best-fit analysis as an instrumental act of vanishing an assumed standard of being within calculation and reason, his work opens further questions concerning the concept of prototypicality. Here, our spatiotemporal position in the world extends beyond the hierarchal classification of being into a differential state of being, where the body of difference is not only fragmented into an alienable neurosis,

14 Pablo Mancosu, *Philosophy of Mathematics & Mathematical Practices in the Seventeenth Century* (Oxford: Oxford University Press, 1996), 173.

as Fanon claims, but also becomes fragmented and valuable only inasmuch as this being can be modulated between operations of accumulation, exploitation, and generalization at the behest of scientific reason. This mode of existence is not merely one of dynamic positioning, but rather a matter of perception as method for empirical decision-making. Modes of nonlinear perception in artificial neural networks might rebuild the relation between the colonial imaginary and the bioepistemic relation. Parisi argues: "Rather than demarcating a simple opposition between theoretical and practical knowledge, [...] current computational practices reconfigure the relationship between reason and automation so that automation is no longer a reflection of a Cartesian dualism of mind and body [...] or of the Laplacian mechanization of the universe."[15]

If we consider Fanon's argument that the problem of Black existence is a problem of perception, as well as the ontological abandonment of Black experience, then the differential can be understood as a device that operates within a relation of bioepistemic ordering and forms of prototypicality. In this way, Leibniz's proposal of a single monadic substance might be rearticulated as a substance the precedes any perception of difference. Consequently, any phenomenon of Black existence, as difference in itself, is predicated on the subsumption of that difference into a uniform "harmony" of relation. If this is so, then how do we account for the immediacy of anti-Black racism within domains of relation? Furthermore, if nonlinear machine learning models, like artificial neural networks, are informed by an ontological neglect of difference—which is materialized in differential calculus—then how does this change the terms of machine learning research? Ultimately, to what extent can Black existence be dislodged from neurosis in domains where Black life is mediated by machine learning algorithms? Furthermore, if race—as Wynter posits—is the single substance in which the Western concept of all being is conceptualized, then the bioepistemic operation of existence is the result of a differentiation

15 Parisi, "Automated Thinking and the Limits of Reason," 474.

between the instantaneous change of one quality of relation (whiteness) relative to another (Black) or the primal quality that differentiates between similar human objects. Framing race as difference can be seen as an act of showing difference within a single species, yet is also an act that involves the containing of one or more derivatives of human origin. Although Wynter situates this origin as a problem of the Western concept of the dis-divination toward the ceremony of white domination, a critical entry point into understanding race opens up to the derived functions of race as set forth in the bioepistemic. In other words, race as a quality of differentiation between similar things is actualized in both the showing of difference as well as the permitted coefficient of a primal way of being and becoming of one human relation relative to another. What are the implications of a techno-social that, as Foucault writes, "simultaneously manifest and obscure empirical order wherever they posit themselves"—where nonlinear relations that exist between bodies of difference or bodies outside of machine perception, or even the drama of the mundane, are valued inasmuch as they can be optimized into discrete empirical perceptions?[16]

The brevity of Leibniz's conundrum is contentious in the ontological study of being. Deleuze finds fault in Leibniz's concrete assumption of difference and change. In his lectures on Leibniz, Deleuze asks us to

> understand that at this level, the notion of compossibility becomes very strange: what is going to make me say that two things are compossible and that two other things are incompossible? […] What is this very unusual relation of compossibility? Understand that perhaps this is the same question as what is infinite analysis, but it does not have the same outline. So we can draw a dream out of it, we can have this dream on several levels. You dream, and a kind of wizard is there who makes you enter a palace; this palace […] it's the dream of Apollodorus told by Leibniz.[17]

---

16 Foucault, *Order of Things*, 51.

17 Gilles Deleuze, "Seminar on Leibniz: Philosophy and the Creation of Concepts" (lecture, April 22, 1980), https://deleuze.cla.purdue.edu/seminars/leibniz-philosophy-and-creation-concepts/lecture-02.

Much of the strangeness in Leibniz's universal theory of existence is relevant to dynamic, generative machine learning algorithms. Machine learning algorithms are, today, powered by derivatives and differentials, which are advanced forms of calculus that can account for nonuniform structures and thus produce more accurate targets. For instance, artificial neural networks are powerful types of nonlinear machine learning modeling. Artificial neural network models are made nonlinear by the sigmoid or activation function. The sigmoid function is used to solve complex (nontrivial) problems, including machine perception. In machine learning, perception can be defined as the capacity to interpret data in a manner that is similar to the way that humans sense and relate to the world. In biologically inspired neural networks, the sigmoid function is usually an abstract representation of the rate of neuronal firing activity, which is typically a binary process. Neural networks are, in this sense, better able to interpret visual input patterns than linear machine learning models. Neural networks serve as the basis of robot vision and autonomous land vehicles. In calculus, the sigmoid is a differential function, meaning it is the derivative of a real variable that exists at each point in a given domain. A derivative is a linear approximation that measures the sensitivity to change of an output value with respect to an input value. A differentiable function must be continuous. As such, a function is said to be continuously differentiable if the derivative is itself a continuous function.

A Correction of Metaphysics and the Concept of the Black Relational Object

It is valuable to consider contemporary machine learning from this brief encounter with theories of linearity and nonlinearity. As the techno-social milieu gains increased abilities to not only capture discrete motion but also simulate the discretization of phenomenon into data-hungry machine learning devices—devices coded to simulate dynamic motion as ideal reductions—the genesis of the human being, being human, is at risk of becoming as obscure as the square root of negative two, a "real" number only inasmuch as it is a necessary representation for instrumentalization and empirical reason.

When does an ideal type of body, a perfection by way of algorithmic accuracy, serve as the governance of individuation? The implications of these "well-founded fictions" merit further scrutiny and investigation. In machine learning, the subject is allowed to come into corresponding relations as a monadic property or accident. As with the Aristotelean substance, when two or more subjects are related, in machine learning their relation can be explained by monadic properties or accidents present in the corresponding data as a natural and concomitant occurrence within the relata. A nuanced approach to machine learning perhaps brings into the debate the position of the subject of inquiry, and more so the subject itself. These mechanizations involve difference as well as motion and computational decision, reducing phenomena such as emotions, motivations, and objectives down to determinant actions that are conditional only in that the conditions can be derived from an assumed point of reference on the shared spatiotemporal coordinate. As such, one's road rage is no longer the metaphysical collision of a bad mood and someone else's bad driving, but nonlinear behavior carried out by both individuals as discreet units that the algorithm must account for through computational decision-making—sometimes at a rate of millions of times per second. In other words, the graphical interface (here, texture mapping) forms an empirical bridge between perception of the real world and analyses of continuously variable quanta.

Having staged the ontological and operational problem of existence as a problematic of nonlinearity and compossibility, we are tasked with questioning whether the subsumption of particulars into generalized patterns might constitute, as Fanon states, "nothing more nor less than *man's surrender*," to a calculus of power.

If Only We Were Able to Set the Black Technical Object Free

If data and nonlinearity within Leibniz's legacy can be conceived of as the simplest, indivisible body within a single substance, then any imagined associative aggregates or incompossibility among them can also be viewed as extensions of preestablished relations that stand in colonial confidence with the universal substance of race. In terms of machine learning, what appears largely as an opposition between human and machine does in effect alert us to specific opportunities to revise our racial perceptions. There is opportunity for authentic engagement in lived experience in recognition of specific historicities. However, there is the potential for a more fully realized being that is expressed from within the duress of a racialized historicity. Being as such, a Black technical being would work through the particular (in this case duress) to emerge as autonomous in its logics of enumerative living that can be articulated as an alternative politics and Black techno-sociality. If only we could set Black life—as well as machine learning technology—free.

Machine learning mediates more than any other technology the conflict of these two poles, although it is machine learning that also attempts to combine them. In this way, machine learning is the combination of the gaze of colonial enumeration and the individual that attaches an indifference to the mechanical gaze that records its actions. Even as machine learning uses this double function of illumination in service of race, it is nevertheless a process that can raise awareness of the capacity to mute the impression of categorization and its normalizing truth-driven forces. This is because although machine learning articulates the "real," it is not bound to our present conceptions of reality or their corresponding racial structures. This gives machine learning, and the Black psyche, leverage to play around with dissonances it can create between the operation of the social and its preemptive interpretations.

The power of contingency emerges in machine learning from its capacity to fail in its operation. The power of error can join with the power of misrecognition of Black life in computation to provide a series of provocations and patterns that, at minimum, distract the colonial paradigm in its obsession with truth. Equally, the Black body retains the right to alter its relation with machine learning by dictating the terms under which data is input into the machine. This does not, however, operate functionally, as data aggregation is powerful and ubiquitous. But it does provide potential for expression that resides in the gaps of existing nontechnical knowledge for the generation of new meaning. This move, however, necessitates a pathological viewpoint that resists the temptations of representation and revision in computational culture. To insert Blackness into the machine is to first assume that the machine does not already understand the incalculability of Black life, which is an affirmative assumption. But it also assumes that returning the mechanic gaze of the master will result in the machine's recognition of its value. This is, of course, an extraordinary success within the pathological view of lack, but it denies any other pathological framework that articulates a totality and self-actualization.

This view assumes a self-centeredness within machine learning, as well as the capacity of data to extrapolate enough of Black sociality to render the Black body both visible and vulnerable to pathological fragmentation. Although the predominant studies of race and technology are aware of the social constructions of race, technology, and the body, they run the risk of "placing the individual into the system of reality," as Gilbert Simondon describes, while "explaining the characteristics of the individual, without a necessary relation to other aspects of being that could be correlatives of the appearance of an individuated reality."[18] Machine learning articulates its own genealogy and computational logics that necessitate a reformulation of human relationships with data. For instance, in terms of neural networks, data is modulated by activation functions. These functions self-regulate their own sensitivities to environmental inputs; data is then reconfigured at the level of the computational premises of prediction (a re-hypothesizing of the premises)

18 Gilbert Simondon, "The Position of the Problem of Ontogenesis," trans. Gregory Flanders, *Parrhesia* 7 (2009): 4.

within the boundaries of the network's specific computational logics. What might appear to be categorical output is instead a dynamic relation that operates prior to discrete correlation and categorical output.

Again, our goal is to understand how the Black technical object operates from within the imposition of categorical output. It is here, at the junction of encounter and experience, that Guattari views the racialized group as assigning meaning to themselves and the world. This meaning is a force that "constitutes the seeds of the production of subjectivity," as "we are not in the presence of a passively representative image, but a vector of subjectivation."[19] It is through the meaning of Blackness that the Black, brown, and racialized individual creates a cohesion of (mis)representation, expounded by aesthetic markers, dynamic vibrations, and cultural kineticisms often expressed as a sense of belonging.

Deleuze asserts that the individual is continually coded and recoded with a moral directive. Here, the Black being becomes the primary agent in individual experiences of duress, understood as a relation of forces. This being also holds the capacity to diagrammatize their own subjective experience. They are, therefore, both internally constituted and regulated. Still, the abstract relation would seem to encourage a negentropic response by moving toward equilibrium, particularly in association with the efficiency of municipal jurisdictions. Negentropy is the measure of a reduction in entropy or increase in order. Negentropy thus gauges a system's perceived normality or stability. Attempting homeostasis implies movement toward system equilibrium, where a system evolves toward a stationary state described by the minimum entropy production. Ilya Prigogine and Isabelle Stengers argue, however, that entropy is constrained by a system's boundary conditions and any observed stability is taken for granted. When a system's thermodynamic forces breach a

19 Félix Guattari, *Chaosmosis: An Ethico-aesthetic Paradigm*, trans. Paul Baines and Julian Pefanis (Bloomington: Indiana University Press, 1995), 25.

threshold of linearity, the system gains sensitivity to fluctuation. The system then reacts (in varying degrees) to these fluctuations by challenging the certainty of order. In some cases, Prigogine and Stengers write, “the analysis leads to the conclusion that the state is ‘unstable.’”[20]

With this ontological starting point, race becomes the technically fragmented Black body, superseding Black relation to duress as techno-resistance. However, this techno-resistance is not in refusal of or in compliance with domination. Techno-resistance is a quantum generation of self, a consistency in and of itself that thrives within as well as in excess of modes of domination. At the same time, the techno-resistant body is a methodological project that makes use of the crisis of the Black body in awareness of network of relations that articulate what we know as sociality. To think through the techno-resistant, however, is to think about contemporary ecologies of generative technical apparatuses such as machine learning, neural networks, and large data systems. What does the Black body mean within this relation, and likewise how does the perceptual articulation of the Black body in terms of Blackness unfold within the circumstances of futurity? This notion requires an altered relationship with the particular and the universal.

As such, the Black technical object invites new multivalent modes of representation. It argues for a nonlinearity as a key point of entry into future techno-human relation. While the Black technical object might be externally positioned outside of ontology, it is in its pre-individuated state the constitution of a potential that exists prior to the amplification of machinic perception or the colonial logics of racial substance. Here, the Black technical object prefigures the constitution of any white prototypicality and instead resides in a continual process of becoming that which is fictively perceived as an imposition of ontological truth. Within this domain, Black existence is not presupposed by oppression or the gaze of whiteness. It is a relation

20 Ilya Prigogine and Isabelle Stengers, *Order Out of Chaos: Man's New Dialogue with Nature* (London: Verso, 2018), 140.

itself, to be manifested in the domain after it has self-actualized. To think through Black being in this way—as an ontogenetic phase of existence prior to the racialized body—requires an alternative type of psychosocial relation that is materialized on unstructured ground. This relation precedes any presupposition or psychic perception of fragmentation or alienation in the spatiotemporal domain. The relation provides a perception of the self that is preconditioned with the capacity to align itself with a totality of being that is mutually constitutive of the lived experience of all humans. This gesture would also require a reading of the senses that challenges the pathological spaces of lack and despair. Instead, the Black technical object develops in affirmation of the present, past, and future self.

Simondon posits that the relation to external objects is a psychical genesis for the individual. This psychical process is parallel and expressive of the operation of individuation. Nonetheless, while the expressive function—like a symbol—references something anterior to individuation, it does not create meaning or signification. Instead, its function is to carry meaning into presence. For Simondon, the individual's existence lies in making actual the process of individuation or the signification of the relation between individuals, while at the same time recognizing that "between individuals there is only signals."[21] In other words, as David Scott notes, "the individual is the phenomenal form signification assumes—it is what being becomes in becoming-individuated."[22] Scott continues: "When Simondon speaks of individualization it is the being of the individual as only ever a problem that seeks resolution, all the more difficult because this incompatibility is internal and, indeed, prompts its own genesis."[23] The individual becomes knowable at the precise moment it acquires phenomenal, or spatiotemporal,

21 Gilbert Simondon, *L'individuation psychique et collective* (Paris: Aubier, 2007), 125, as quoted in David Scott, *Gilbert Simondon's Psychic and Collective Individuation: A Critical Introduction and Guide* (Edinburgh: Edinburgh University Press, 2014), 94.

22 Scott, 94.

23 Scott, 94–95.

being. The individual's genesis corresponds to the resolution of a problem that cannot be resolved as a function of anterior givens that do not have a common axiomatic: "*the individual is the auto-constitution of a topology of being that resolves an anterior incompatibility through the appearing of a new systematic.*"[24]

Simondon's ontology of psychic individuation further complicates the genesis of the Black technical object in its confrontation with the substance of race. He finds it problematic to approach the individual as the point of departure. The consequence is that if we start to think about the racial being from the point of view of the individual, then individuation is reduced to mere representation. Alternately, to reform our categories of relation and knowledge requires a radical reconception of what an individual is, as opposed to that individual being relegated to the process of being racialized. This unity of internal (psychic) and external (collective) process of individuation is justified in the examination of the ontogenic conditions and circumstances that have solidified into normative social values. In other words, there is a case for understanding the psychosocial conditions of racialized being as life is extracted and abstracted into normative social structures. Simondon writes:

> We cannot say that the concept of the "pathological" is the logical contradictory of the concept "normal," for life in the pathological state is not the absence of norms but the presence of other norms. Rigorously speaking, "pathological" is the vital contrary of "healthy" and not the logical contradictory of "normal." [...] Disease—the pathological state—is not the loss of norm but the aspect of a life regulated by norms that are vitally inferior or depreciated, insofar as they prevent the living being from an active and comfortable participation, generative of confidence and assurance, in the kind of life previously belonging to it and still permitted to others.[25]

The Black technical object must first be considered, prototypically, as a communion of generative individuals, as opposed to a

24 Simondon in Scott, 95.
25 Simondon in Scott, 109.

static prototype of whiteness. Additionally, the process of becoming Black and the aesthetics of Blackness must be an operational mode of living. The individual residing in a preexisting milieu of cultural forces harbors an incompatibility with themselves and their surroundings. Racial beings hold the capacity for ontogenesis or a self-creation within systems of identification and this self-production assumes an independent, do-it-yourself status. Furthermore, duress is an attractive force for collective consistency through which the latent potential of Black people can be articulated. Simondon argues: "In a theory of the phases of being, becoming is something other than an alteration or a succession of states comparable to a serial development. Becoming is in effect a perpetuated and renovated resolution, an incorporating resolution, proceeding through crises, such that its sense is in its center, not at its origin or its end."[26]

Following Simondon's development of the process of individuation, the individual is always incomplete and finds themselves continually involved in new processes of becoming. The individual already belongs to anything that contributes to establishing relations, including relations with oneself. Simondon moves a step further by theorizing that while in the process of becoming, the individual is also involved in an operation by which other individuals form consistencies of connection. This transindividual comes into view from successive operations of individuation. It is within the framework that Black peoples and Black culture exist as consistencies of relation. The emergence of Blackness, therefore, articulates more than a group of people unified under a common feeling of discontent. Instead, Black existence is developed through the process of being and becoming Black.

Luce Irigaray suggests that such a provocation would require a sharing of the social space that encourages the projection of a different human reality where one sees the other as

26 Gilbert Simondon, "The Genesis of the Individual," in *Incorporations*, ed. Jonathan Crary and Sanford Kwinter (New York: Zone Books, 1992), 223.

irreducible to one's own experience. Irigaray argues that this world is constituted through "respecting the other within oneself, and not by projecting the totality of what exists outside of oneself"—as systems of domination often do.[27] She pushes for a conviviality in excess of a passive acceptance of "difference" that has failed to articulate itself within mathematical reason. In this way, the core of the social is revealed in its most explicit realization: that the conditions of human reality are irreducible to the relation of the other. Being and becoming would then depend on our ability to achieve "another world, another nature, and not on our transforming this world and our own nature."[28] They would also depend on the circumvention of reactionary postures by the operations of data that paralyze social relation. Simply put, the authenticity of the social would have to become an authenticity that is difference in itself.

We must look beyond racial representation as well as technological function to think through the generation of Black being and psychic development. Here, we might consider the substance of race as already present in the techno-human relation. The seeming polarity between humans and data is symptomatic of a larger overlapping concern about the origin of Black political subjectivity and the trauma of colonialism and slavery. As such, the overriding problematic rests in the presupposition of negation in reflection of the master. The manner in which the problem of race and the technical apparatus is framed results from principles that begin with an observation of being that is anterior to the conditions that bring the individual into existence. This frame not only situates the Black body in its present concreteness, but forecloses any turbulent iterations of being that have come to inform the network of relations between the Black body and its environment. To do otherwise would move Blackness toward a different sense of desire. This desire would pull away from a longing for the non-Black Other that Fanon

27 Luce Irigaray, *Sharing the World: From Intimate to Global Relations* (London: Continuum, 2008), xiv.

28 Irigaray, xvii.

proposes; instead, it is substituted for one that gains capacity in a return to its nonessentialist origins.

It is this point of ontological departure that makes the building of a new future plausible. If the Black body does enter into a space of the imaginary to work through the immediacy of race and racialization, then the future is collapsed into the present, not as a denial of hyperstition, but as an always already pre-individuated potential for the generation of subjective futurity. Furthermore, the impasse is one that does not discard Fanon, but works through the foundations of Black representation to formulate a more self-affirmative and optimistic view of Black life, or what Moten calls a pre-op(tical) optimism in Fanon. Moten argues that "blackness has been associated with a certain sense of decay, even when that decay is invoked in the name of a certain (fetishization of) vitality."[29] Consequently, as Moten argues, Black life is driven toward a being-toward-death or a death-driven nonbeing within institutional discourse, as opposed to a more complete assessment of the impact of Black psychic and collective generation, which is produced at the "barbershop, the beauty salon, and the bookstore" and other important spaces where Black culture is developed.[30] In other words, what we call Black, the Black body, and Blackness is more closely aligned with the languages, discourses, debates, and creative projects found in the everyday making of Black life than in the halls of academic and cultural institutions. Blackness is far more than the negation of the Black experience under colonialism and racism, which is why Moten has fervently asked, "If, as Frantz Fanon suggests, the black cannot be an other for another Black, if the black can only be an other for a white, then is there ever anything called black social life?"[31] Can the Black technical object build an awareness toward its existence beyond the negating principles of decay? Can Black life instead be viewed as generative, affirmative, and fundamental to a technically mediated life?

29 Moten, "The Case of Blackness," 177.
30 Moten, 178.
31 Moten, 178.

Here we might consider the Black technical object as a distortion of the psychic effects of patterned, modeled, and calculable alienation as it assumes an internal indifference to the substance of race inasmuch as it produces a relation with the technical as an act of thought in machinic process. The risk, which Fanon describes as an upsurge of transformation in the face despair, opens up the shadowy and often ignored symptoms of the techno-human relation to the constitution of a new and much more expansive transindividual relation. This relation is mutually constitutive of the individual and collective generation of psycho-technical awareness that, as Foucault eloquently claims, might "tame the wild profusion of existing things [in] our age-old distinction between the Same [race] and Other [technical object]."[32] By seeking to arrive at this psycho-technical awareness, we are invited to reflect on what it might mean to challenge the presupposition of the individual whereby Black existence, as a mode of substantialist metaphysics, is grasped from the perspective of the substance of race. Instead, we march toward an embrace of the always already incomplete and incompatible process of racial and machinic individuation. Given the aforementioned provocation, let us think through what the problem of Black individuation might look like within an ontology of the algorithmic, and draw attention to an alternative mode of Black existence. I return to Buolamwini's imagining of the future and the hall of possibilities, where the shape-shifting human-spider Anansi inspires us to transform our apparent weaknesses into virtues, our linear and substantial representations into cunning and creativity, and our ancestral wisdom into awareness of our pre-individual capacity for survival within the wicked web of incompossible data.

32 Foucault, *Order of Things*, xvi.

This is all I have to give.

These words appear as the acknowledgments in Seal's third studio release, *Human Being* (1998), arguably the most raw and honest album of the recording artist's oeuvre. Over the years, I have understood this emphatic declaration as an endowment of sorts—an affirmative edict, if you will, of how one might live, act, breathe, be in a world that calls for us to care for ourselves and one another, even as the evidence seems to suggests otherwise. A line has been drawn. Some would say this line has always existed, whereby we are led to believe that the expression of our authentic selves in this strenuous and often violent world is as foolish as it is futile. Whereby an act of thought or of aspiration is of no practical use in a world hurried to solve problems not entirely understood. Meanwhile, the most disparaged in this world live in desperation as self-interest is substituted for compassion, profit for care, and discipline for nourishment. After all, as Seal proclaims: "In my hour of desperate need … They'll be punchin' tickets by the minute if you fall out of line." How do we create a world otherwise?

I have had my ticket punched more times than imaginable in my refusal to fall in line. In my *many* hours of desperate need, when the punch seemed so hard that I had all but lost faith in myself and everything beautiful in this world, I have been reminded by those who love me that to be human is to bleed. But to be human is also to love. In fact, these folks have shown the immense courage, wisdom, and compassion it takes to give oneself over fully to love. They did so even when they too believed it was all they had to give, when they felt the troubles of this earth were so big that they seemed to block the entire sun. But they loved anyway, and in spite of everything. They taught me how to love. And for this, I am eternally grateful.

To my mentor, Daisaku Ikeda, who taught me to dream of beauty in a world full of nightmares. I am gracious for your belief in me and undying support. My sincerest thanks to Professor Matthew Fuller and Professor Luciana Parisi, who, through their patience, rigor, and kindness, saw in me what at times I refused to see in myself. To Professor Ezekiel Dixon-Román for his generosity, care, and support. To my tribe: Dr. Nicole

Sansone Ruiz, Dr. Mijke van der Drift, Dr. Chana Morgenstern, and GiGi Demming; I treasure the intellectual debates, exchanges of laughter, and limitless compassion you offer this world. To Niels Schrader, Dr. Sheena Calvert, Dr. Marina Otero Verzier, Klaas Kuitenbrouwer, and the wonderful community at Royal Academy of Art, The Hague, and Het Nieuwe Instituut, Rotterdam, thank you for the space you gave me to explore many of the themes in this book. And thank you for your tireless efforts entertaining and supporting my wildest ideas.

To my partner and joy, Axel Satgé. This book would not have been possible without you. I cannot imagine a world without you by my side. Your time, energy, love, and encouragement will forever live in my heart, and in this text.

Finally, I would like to acknowledge my loving family, particularly my parents. When I was young, my father, E. Samuel Johnson III, was asked to write a letter to describe me. I do not remember it all, but I vividly recall his struggle with the task. How could he put into words what he saw when he looked at me? He wrote that to ask how much he loved me would be like asking a grain of sugar how sweet it is, an ocean how wet, or a sky how blue. This is a love we share.

And to my dear mother, Phyllis Teresa Johnson. How lucky I am to have had you in my life. You were always full of wisdom and infinite amounts of joy. You taught me how to love. You taught me what courage is. You gave me everything—all that you had to give. I only hope, during your brief time in this world, that I was able to show just how much you meant to me. I dedicate this work to you.

For now, this is all I have to give.

Ramon Amaro
The Black Technical Object
On Machine Learning and the Aspiration of Black Being

Published by Sternberg Press

On the Antipolitical, volume 2
Series Editor: Ana Teixeira Pinto

Copyediting: Leah Whitman-Salkin
Proofreading: Raphael Wolf
Design: Joaquín Gáñez
Printer: Tallinn Book Printers

ISBN 978-3-95679-563-3

Distributed by The MIT Press, Art Data, Les presses du réel, and Idea Books

Sternberg Press
71–75 Shelton Street
London WC2H 9JQ
www.sternberg-press.com